# Chemistry at Home
## Exploring the Ingredients in Everyday

# Chemistry at Home
## Exploring the Ingredients in Everyday Products

John Emsley

THE QUEEN'S AWARDS
FOR ENTERPRISE:
INTERNATIONAL TRADE
2013

ISBN: 978-1-84973-940-5

A catalogue record for this book is available from the British Library

Published by The Royal Society of Chemistry,
Thomas Graham House, Science Park, Milton Road,
Cambridge CB4 0WF, UK

Registered Charity Number 207890

Visit our website at www.rsc.org/books

Printed and bound by CPI Group (UK) Ltd, Croydon, CR0 4YY

# Disclaimers

The Royal Society of Chemistry is not responsible for individual opinions expressed in this work and does not endorse or recommend the products mentioned herein. Other products are also available. Products should be used in accordance with the manufacturer's instructions.

Readers should be aware that the products discussed in this book were those that were available in the UK in 2014. The ingredients in current versions of these items may be different.

Efforts have been made to locate owners of all reproduced material and we trust that no copyrights have been inadvertently infringed.

# Preface

When spring comes, Nature gets to work making millions of different chemicals. Some are beneficial and some are dangerous, some smell nice and some smell awful, some heal and some harm, some are essential and some are superfluous. Should chemists add other chemicals to this collection? Some people think not, but most people accept that what we produce makes our lives better. However, we must make sure they are not a threat to our health, nor must they pollute the environment when we dispose of them.

If you are one of those people who think that today we are exposed to far too many chemicals, and that these are all suspect, then this is not the book for you. However, if you appreciate that chemicals are what makes life healthier, cleaner, longer, colourful, and safer, then read on. You will discover in *Chemistry at Home* just what these substances are and what they do for us. If you are worried that they might still present a risk to health, then be reassured that testing is now so thorough, and approval for use so difficult to obtain, that nothing is included that might pose a risk to anyone except a few rare individuals. They will be aware for medical reasons that certain perfumes or food ingredients, for example, may affect them and to avoid products that indicate these on their labels. For the rest of us, modern household products are safe to use and they create a home that we can take pleasure in living in.

Chemistry at Home: Exploring the Ingredients in Everyday Products
By John Emsley
© John Emsley, 2015
Published by the Royal Society of Chemistry, www.rsc.org

So what the ingredients in the things we buy and use? Their names mean little or nothing to most people, and that's the reason for writing this book. The ingredients listed on a product label provide information that needs interpreting if we are to understand why they have been included. Sadly for chemists, some people have done this interpreting already and supplied the media and internet with their opinions. Of course there have been chemicals used in the past that we now know should not have been used, but that was at a time when health and safety testing was in its infancy. Today things are very different, but there are still those who are inclined to condemn almost every-thing that chemists produce, the implication being that they are unnatural and so must be harmful.

Walk round a supermarket, a garden centre, or DIY store, and you are surrounded by thousands of different products com-peting for your attention, although many are very similar.

Want a household cleaner? Here are 20 to choose from, some being variants of the same brand. Want a shampoo? There are scores of those, every one claiming a special benefit. Want something refreshing to drink that doesn't add calories to your diet? Lots to choose from there. Of course many of the same kinds of product contain the same kinds of ingredients, and in this book I have chosen a representative selection. They are typical of those you will find in most UK homes, and I have chosen to discuss them in chapters devoted to the various locations where you might encounter them.

The main entry for each category will explain what a typical product is designed to do and the chemistry involved in its use. Then follows a list of the ingredients and I give the reason for their being included in the product. Often I will be repeating the same information and explanation, but I have done this because *Chemistry at Home* is a book to be read as a source of information and you may only need the information on one particular page. For those who wish to know more, there is a Glossary in which the ingredient names are translated into chemical names and formulae, together with their reason for being used.

The major manufacturers generally list all the ingredients on their products or on their websites. In some cases, such as cos-metics, there may be more than 20 – often in tiny lettering making it difficult to read – which to some may look a little suspicious. I suspect most consumers ignore the list, trusting as they do that all chemicals have been tested as safe – as indeed they are.

This book is published by the Royal Society of Chemistry. On its LearnChemistry website is a section called 'Chemistry in your cupboard,' which includes nine products: Calgon®, Cillit Bang®, Dettol®, Finish®, Harpic®, Gaviscon®, Nurofen®, Vanish®, and Veet®, all of which are products of the Reckitt Benckiser® company. Some of these are also in this book, but here we look at them slightly differently, concentrating on saying exactly what every ingredient is and what it is there to do.

There are still some ingredients that are targeted by groups who run campaigns against them, although often their claims are not backed up by meaningful research and tend simply to be scare stories. They like to report finding man-made chemicals in unexpected places, such as in blood – the implication that these pose a serious risk. What they fail to explain is that the amounts detected are in parts per billion (ppb), a unit unfamiliar to most people. However, 1 ppb is like 1 second in 30 years – not a lot. Nevertheless, the result of such scare stories may be that companies whose products include these chemicals will replace them because they are fearful of the adverse publicity that has been generated. However, we still need many ingredients that these groups campaign against, such as sweeteners, preservatives, disinfectants, pesticides, and fragrances, which may have been used for years because of the benefits they confer.

The Sense About Science movement has published a booklet called *Making Sense of Chemical Stories* and a new edition of this came out in 2014. It got very sympathetic media coverage, and I would like to think that a rational debate can now begin of the role of chemicals in everyday life. All being well, the days of ill-informed suspicion about the work of chemists are coming to an end.

I've said enough. So start reading and discover what's really in the things you buy. And remember that products are continually being launched and existing ones re-formulated or even removed from the market place. What you read about in this book relates to the state of play in 2014.

John Emsley

# Acknowledgements

First let me acknowledge those unsung heroes of the modern world: the chemists who have provided us with all the products that are in *Chemistry at Home*: food chemists, fragrance chemists, analytical chemists, pharmaceutical chemists, cosmetic chemists, colour chemists, horticultural chemists, polymer chemists, and research chemists. Without their skills we would be living as people lived 200 years ago, in a world where for the vast majority of people life was short, hungry, dangerous, dirty, dull, and painful. Chemistry is the science that transformed life for everyone in developed countries and will, I hope, one day do the same for all people in the world.

I also need to acknowledge more direct help from others. Although I understand the chemicals used as ingredients in terms of their molecular structure and properties, when it came to explaining why these were used in the many products discussed in this book I was in less familiar territory, although not entirely so. For many years I was the chemist consultant for the Broadcast Advertising Clearance Centre – now called Clearcast – whose job it is to see that claims made in television adverts are supported by scientific evidence. My role was to check adverts for household products which involved chemicals, such as cleaners, detergents, stain removers, air fresheners, and the like. This entailed visiting the research laboratories of the various manufacturers to see the evidence, which meant inspecting their

Chemistry at Home: Exploring the Ingredients in Everyday Products
By John Emsley
© John Emsley, 2015
Published by the Royal Society of Chemistry, www.rsc.org

xi

research facilities and meeting their research chemists. Invariably the labs were well funded and their researchers were qualified chemists. Customer care was a key factor in what they did.

Having written the chapters of *Chemistry at Home,* I needed to consult experts and ask them to check what I had written. Thankfully I was able to count on the help of various individuals whom I have met over the years. One particularly helpful one was Sir Geoffrey Allen, one time Head of Research of the international consumer products company, Unilever, and he put me in touch with John Russell, retired chemical engineer who came up with some useful suggestions. Others who were prevailed upon to read what I'd written were the following:

Alistair Crawford MD. He diagnosed Chapter 1 and suggested remedies.

Chapters 2 and 5 were laundered by a specialist in detergents who removed any stains.

Dr Christopher Flower of the Cosmetic, Toiletry & Perfumery Association examined Chapter 7 and advised how it could be made more appealing.

Gem Bektas, of the Society of Cosmetic Scientists, examined Chapters 3 and 7.

For Chapters 4 and 6, I take full responsibility.

Chapter 2 was also inspected by Colin Butler, Regulatory Affairs Manager at Reckitt Benckiser and Stuart Bell, Reckitt Benckiser who cleaned it up.

The chapters devoted to food, its preparation and consumption, namely 8, 9, and 10, were chewed over by Dr Sandy Lawrie, a long-time employee of the UK Government's Food Standards Agency, and then the Food Standards Agency, who improved several of them.

Philip Glynn-Davies of Jaguar Land Rover drove through Chapter 11 and pronounced it free of obstacles.

Susan McGrath-Cole, a member of the Society of Chemical Industry's Horticulture section, inspected Chapter 12 and found nothing to report.

Drs Marshall and Mary Smalley examined Chapter 1 and weeded Chapter 12 and came up with cures for the former and suggestions for improving the look of latter.

Catherine Adams read Chapters 1 and 2 and made useful suggestions.

Professor Steve and Rose Ley of Cambridge were persuaded to read all of the manuscript and made invaluable comments.

And of course my wife, Joan, who purchased many of the products discussed in this book and offered useful comments about their efficacy.

Finally I should like to thank Kathryn Duncan for editing my text thoroughly and conscientiously, and uncovering several inconsistencies that have now been corrected; and Sylvia Pegg, who is Senior Production Controller for Books at the Royal Society of Chemistry.

# Biography

John Emsley has written 12 popular science books over the past 20 years and several of them have been about the impact of chemistry in everyday life. His first was *Consumer's Good Chemical Guide,* which appeared in 1994 and won the Science Book Prize. Other followed such as *Molecules at an Exhibition* in 1998, *Vanity, Vitality, Virility* in 2004, which is about cosmetics, food and sex, and *Better Looking, Better Living, Better Loving* in 2007, which continued those key themes. *A Healthy, Wealthy, Sustainable World* appeared in 2010 and in this he discussed the products we have come to rely on, in terms of whether they were sustainable. More information about the author can be found on www.johnemsley.com.

Chemistry at Home: Exploring the Ingredients in Everyday Products
By John Emsley
© John Emsley, 2015
Published by the Royal Society of Chemistry, www.rsc.org

# Contents

Chemistry at Home: Exploring the Ingredients in Everyday Products
By John Emsley
© John Emsley, 2015
Published by the Royal Society of Chemistry, www.rsc.org

# Technical Words You May Need to Help You Understand the Text

There are some terms in *Chemistry at Home* that you might be unfamiliar with but which categorise certain requirements that a product must have if it is to deliver the results we are seeking. One you might be familiar with, but which has been much misunderstood and even maligned in the past, is the system of giving E numbers to food additives. This was introduced as a way of reassuring consumers that something was safe. Let me explain.

**E numbers** are allocated to ingredients that are used in foods, certifying them as safe. These are only given an E code when they are permitted to be used in all countries of the EU and approval is given by the European Food Safety Authority (EFSA).

Although E numbers were meant to reassure consumers that ingredients were perfectly safe, they came under a sustained media attack from some vested interest groups who believe that all man-made chemicals are inherently harmful. They were supported in this by Maurice Hanssen in his 1984 book, *E for Additives*, which became a best-seller despite being ill-informed and alarmist in tone. The Royal Society of Chemistry has many members who work for food companies and it publishes a proper and scientific account of the subject called *Essential Guide to Food Additives*, which is now in its 4th edition.

Chemistry at Home: Exploring the Ingredients in Everyday Products
By John Emsley
© John Emsley, 2015
Published by the Royal Society of Chemistry, www.rsc.org

Here are other terms that you might be unfamiliar with, listed in alphabetical order:

**Aqua** occurs in many lists of ingredients and this is simply the Latin word for water, which is almost always present as a solvent.

**Antimicrobial agents** are there to kill any microbes (germs) that might contaminate food or a personal care product. Microbes like bacteria and moulds can cause illness. When the thing to be preserved is never going to be consumed, then more powerful agents can be employed, such as those that release formaldehyde, which is deadly to microbes. These agents are imidazolidinyl urea, DMDM hydantoin, quaternium-15, and diazolidinyl urea.

   Other highly effective antimicrobial agents are derivatives of benzoic acid, such as sodium benzoate, and of *para*-hydroxybenzoic acid such as the methyl, ethyl, and propyl esters, collectively known as parabens. Some of these occur naturally. These have the advantage of being stable over a range of pHs (acids and alkalis). Sorbic acid and lactic acid are also antimicrobial agents, including their derivatives such as potassium sorbate, which is added to low fat spread. They are approved for use in foods and so have been assigned E numbers.

**Antioxidants** are needed to protect our body against naturally-generated free radicals, which are essentially rogue chemicals. Those free radicals that involve oxygen are the most dangerous and we have a natural system of protection within us to deal with them. The products that we use also need protection, especially if they contain unsaturated fats and oils which are susceptible to oxidation, which makes them rancid. Some antioxidants are vitamins – namely vitamins C and E, which are water-soluble and fat-soluble, respectively. Others are butylated hydroxytoluene (BHT) and butylated hydroxyanisole (BHA). Gallic acid and sodium gallate are also good antioxidants. Those are approved for use in foods in the EU and have E numbers.

**Buffering agent**. This is designed to keep the pH of a product stable at the value at which it works best. Some are designed to keep a liquid acid, pH less than 7, and some to keep it alkaline, pH greater than 7. Buffers are generally weak acids or weak bases.

**Chelating agent (chelant)** or **sequestering agent**. This is a chemical which traps unwanted metal ions, and it does this by wrapping itself around them and thereby disabling them. Hard water is caused by calcium and magnesium ions and they can interfere with ingredients in all kinds of products such as those used in washing. For example, they will form a scum when soap is used. Typical chelants are monoethanolamime citrate and trimonoethanolamine etidronate.

**Emollient**. This is something which softens the skin and is added to personal care products such as shampoos, shower gels, and cosmetics. It may also act as a moisturiser to prevent the skin becoming too dry. Silicones are often used, although for very dry skin a hydrocarbon like Vaseline® may be better. You will read about several of these in Chapter 3 (The Bathroom) and Chapter 7 (The Bedroom). Typical ones are cetyl alcohol and methicone which perform this service in many products, especially beauty creams.

**Emulsifier/stabiliser**. When a food needs to be a smooth blend of oil and water, as in a mayonnaise or low calorie margarine, then it requires an ingredient to ensure these remain blended and don't separate into two layers. Lecithin, ammonium phosphatide, and the glycerides of fatty acids perform this service. You will meet these when you visit Chapter 9 (The Dining Room). Some personal care products contain ingredients which would normally not blend together and they too require an emulsifier. You will meet these in Chapter 3, The Bathroom.

**Excipient**. This is something that has been added to a pharmaceutical but which is not an active agent in itself but is there to help. There are several types of excipients: a binder will glue the ingredients together into tablet form, another might be needed to provide a stable outer coating for a capsule. There may be something added to ensure that tablets do not stick to the equipment making them. Colours are generally not necessary for the safe operation of a pill but they do enable it to be recognised. When a medicament is bitter tasting, then sweeteners may be used to make it palatable. For liquid medicines, the excipients can include things to stabilise the pH, or to keep it germ-free.

**Fixative**. This is a chemical that is used along with perfume ingredients to prevent the loss by evaporation of fragrance molecules, thereby extending the life of the product. A commonly used one is benzyl salicylate.

**Gel**. This is a solid material which can be very soft like a jelly when it contains a lot of water, or quite hard like some sweets when it contains very little. Gels exist in this state because of weak cross-linking between water-soluble polymers, which can consist of proteins or carbohydrates.

**Humectant**. This is designed to attract moisture from the atmosphere, or to stop water loss to the atmosphere, and may be added to prevent a product from drying out or may be used in personal care products that are applied to the skin to keep it rehydrated and looking firmer. Typical humectants are propylene and butylene glycols, glycerol (aka glycerine), sorbitol, and the natural chemical aloe vera. What these molecules have are lots of hydroxyl groups (OH) and these enable them to bond to water. Humectants are added to foods to increase shelf-life and to personal care products such as conditioners, skin creams, and toothpaste.

**Mordant**. This is the linking chemical by which a colour molecule becomes attached to its target. Mordants are usually metal salts in which the metal ion has a positive charge of 2 or more. Sometimes the dye and mordant are combined separately before dyeing and are then known as a **lake**.

**Parfum**. Perfumes are a complex mixture of chemicals, some natural, some artificial, and perfume raw material may contain items that some individuals are sensitive to. These will then need to be identified on the label.

**Preservative**. This is a general term referring to something which is added to all kinds of products to extend their shelf-life, or their lifespan when in the home and they have been partly used. If something is contaminated with microbes it might cause illnesses, or if affected by oxidation it may become malodorous. Preservatives kill microbes and are needed not only for foods but for all personal care products like shower gel, toothpaste, moisturiser, make-up, and skin cream. Once these are opened and exposed to the air, microbes will get in and they could start to ferment. If they contain molecules that are susceptible to oxidation, they may begin to react with oxygen from the air. Some preservatives need to be soluble in water, some in oils, and some need to be safe to consume if they are protecting foods.

Linking man-made preservatives to human health issues has made people suspicious of them although there is no convincing proof that they are dangerous to humans. However, manufacturers have become wary of using them and now seek preservatives extracted from natural sources. These may be less effective and in some cases have led to product recalls.

**Propellants**. These are the gases which provide the pressure that ejects a product from its container in the form of a spray. These are often hydrocarbons such as propane and butane or may even be nitrogen gas.

**Surfactant**. An older word for this was detergent, and before that they were simply described as soaps. Surfactant is short for surface-active-agent and this kind of chemical has a water-seeking head and an oil-seeking tail. Surfactants enable oils and grease to mix with water, and in a cleaning product they are there to drag the grease off a fabric or surface (and the dirt that sticks to it) and into the water so it can be washed away.

Some surfactants have a positively charged water-seeking end, some a negatively charged one; some have both positive and negative components, and some carry no charge at all and are neutral. The chemical terms for these are cationic (positive), anionic (negative), amphoteric (both positive and negative), and non-ionic (neutral).

**Viscosity improvers (thickeners)**. These are needed to stiffen a liquid and are used when a product needs to be a gel or a cream but whose natural ingredients would all be runny liquids, even when they have been converted to an emulsion. Some natural gums make effective viscosity improvers.

**Water softener**. This is a chemical designed to remove calcium and magnesium ions from water, and this can be done by chelating them (see above) or trapping them in a structure like a zeolite.

## CALORIES

As a chemist I should talk about energy in units of kilojoules (kJ) and indeed the energy content given on food packaging also uses this unit. However, most people talk of *calories*, by which they really mean kilocalories (kcal) (which are 1000 times larger than calories). The calorie is also cited on food packaging and people know what it means so I have decided to go with the flow and use it. I apologise to my chemist readers, who I hope will understand my reason for doing this.

## EXTRA INFORMATION

If you want to check on the safety of a product's ingredients, then you need to go to its Material Safety Data Sheet (MSDS), which will be accessible from the company website and which by law has to declare any ingredients that might present a health and safety risk and say what action to take if there is an accident involving them. However, these refer to chemicals in their pure state, whereas in products they are at such tiny levels that any risk is equally tiny. Another useful site is the International Nomenclature of Cosmetic Ingredients (INCI), which provides data about all products which come into contact with the human body. There are also several compilations of information such as The Merck Index®.[†] This is a technical manual with information about a compound's alternative chemical names, when it was first announced, its chemical formula, its molecular structure, and the uses to which it is put. And of course there is Wikipedia, whose information is generally reliable.

For cosmetic products, European law (EC regulation no. 1223/2009 of the European Parliament and of the Council 30 November 2009) requires each cosmetic to be individually assessed for safety in use before it may be sold to the public. That assessment has to be done by qualified experts.

---

[†]The name THE MERCK INDEX is owned by Merck Sharp & Dohme Corp., a subsidiary of Merck & Co., Inc., Whitehouse Station, N.J., U.S.A., and is licensed to The Royal Society of Chemistry for use in the U.S.A. and Canada.

# CHAPTER 1

# The Medicine Cabinet

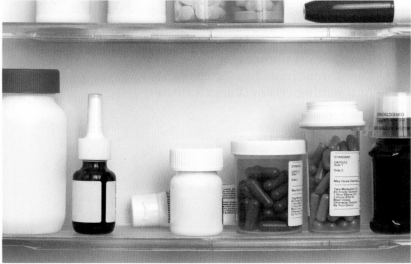

Every home should have a medicine cabinet, which needs to be out of the reach of children. In it are kept the medicines for the various everyday afflictions that beset us, such as headaches, indigestion, coughs, colds, aches and pains, spots, cuts, sore eyes, blocked ears, warts, and other more personal problems. (Constipation relief is covered in Chapter 5.)

---

Chemistry at Home: Exploring the Ingredients in Everyday Products
By John Emsley
© John Emsley, 2015
Published by the Royal Society of Chemistry, www.rsc.org

The family medicine cabinet might also include medication that has to be prescribed by a doctor, and while these are also products produced by pharmaceutical chemists, they are not something included here. The products in our medicine chest are:

1. Pain relief (Solpadeine® Plus, Cuprofen®, Disprin®)
2. Indigestion (Rennie® Spearmint, Alka-Seltzer®, Zantac®)
3. Eye wash (Optrex®)
4. Cough medicine (Venos® Chesty Cough Expectorant, Tixylix Dry Cough Linctus®)
5. Earwax remover (Otex®)
6. Cracked soles of feet (Flexitol® Heel Balm)
7. Rheumatism avoidance and relief (Boots Glucosamine Sulphate, Ibuleve™ Gel, Voltarol®)
8. Skin infections (Bazuka™ Gel, Canesten®)
9. Antiseptic ointment (Savlon®)
10. Diarrhoea (Imodium®)
11. Sore throat remedies (Strepsils®)
12. Anti-itching cream (Eurax® Cream)

## 1.  PAIN RELIEF

*The Chemistry*

There is no need today to suffer the everyday aches and pains that afflicted past generations. There are three popular pain-killing chemicals that can be obtained over-the-counter and these are aspirin, paracetamol, and ibuprofen. Ibuprofen is a good anti-inflammatory, paracetamol is a good painkiller, while aspirin can do both but it brings the risk of stomach bleeding. These medications are also able to reduce fever, the higher temperature being a sign the immune system is fighting infection.

The best and quickest cure for headaches is paracetamol, which comes in various guises, and the one described here contains two other ingredients that can boost its performance.

The best cure for muscle and joint pain is ibuprofen, which suppresses the excess arachidonic acid that is the cause. (We need this molecule to make prostaglandins, hormones, and cell membranes.) In response to a strained muscle, an infection, or an arthritic joint, the body will produce a local excess of prostaglandins and this causes inflammation and pain. Ibuprofen blocks the enzyme that is responsible.

© Shutterstock

Aspirin is recommended as immediate treatment for someone who is suffering a stroke or heart attack because it restores blood flow by thinning the blood, which it does by controlling the cyclo-oxygenase (COX) enzymes. It can even save a life. (Older people should carry a couple of aspirin tablets with them at all times.) Aspirin counteracts the production of platelets in the blood that are designed to stop bleeding but can cause a blockage if they clump together, leading to a stroke or heart attack. For this reason, many people take a small 75 mg tablet of aspirin every day, to prevent this from happening.

There are two kinds of COX enzymes: COX1 protects the lining of the stomach and intestines, whereas COX2 triggers the release of prostaglandins that result in inflammation and pain. Ideally a painkiller should only target COX2, but aspirin targets both.

**Solpadeine® Plus**

This is based on paracetamol and it acts rapidly because it can be absorbed through the stomach wall. When combined with a little codeine and caffeine it is quicker to act and its effects are longer lasting, but this carries a low risk of addiction if used over long periods because codeine is an opioid drug like morphine and heroin. (Overuse of paracetamol is also not recommended because of potential liver damage.)

*Contents*

Paracetamol
Codeine phosphate hemihydrate
Caffeine
Maize starch pre-gelatinised
Potassium sorbate
Povidone

Purified talc
Magnesium stearate
Microcrystalline cellulose
Stearic acid
Hypromellose
Carmoisine

*Active agents*

The amounts of these are: **paracetamol** 500 mg, **codeine phosphate hemihydrate** 8 mg, and **caffeine** 30 mg.

**Maize starch pre-gelatinised** is starch extracted from maize that has been heated to unravel its polymer molecules, making it easier to digest, thereby causing the tablet to break up and release its active ingredients.

**Potassium sorbate** is a preservative that protects the tablet against moulds and yeasts.

**Povidone** is a water-soluble polymer that acts as a glue to hold the ingredients in the tablet together.

**Purified talc** is magnesium silicate and is a soft mineral that breaks up easily. Here it is added to bulk out the tablets.

**Magnesium stearate** and **stearic acid** prevent the tablets from sticking to the equipment that compresses them into tablet form.

**Microcrystalline cellulose** is another bulking agent.

**Hypromellose** is a modified form of cellulose and it acts partly as a glue and partly as a controlled release agent for the active ingredients so that their effects will last longer.

**Carmoisine** is a red colourant.

There are other brands based on paracetamol, such as Panadol® and Anadin™, as well as pharmacist and supermarket own-brands.

## Cuprofen®

In this product the pain relief comes from the action of ibuprofen, which is best for aches and pains rather than headaches.

*Contents*

| | |
|---|---|
| Ibuprofen (400 mg) | Tablet coating is |
| Lactose |   Methyl cellulose (Methocel® E15) |
| Ac-di-sol® |   Polyethylene glycol 500 |
| Methyl cellulose (Methocel® A4D) |   Mastercote® pink FA0430 |
| Magnesium stearate | |

**Ibuprofen (400 mg)** is the active agent.

**Lactose** is also known as milk sugar and is added as filler to bulk out the tablet.

**Ac-di-sol** is sodium croscarmellose, which causes the tablet to swell and rapidly release the ibuprofen in the stomach.

**Methyl cellulose** binds the ingredients together. (Methocel® A4D is the trade name of a particular brand.)

**Magnesium stearate** prevents the ingredients sticking to the machinery that stamps them into tablets.

*Tablet coating*

**Methyl cellulose** holds the coating together. (Methocel® E15 is
the trade name of a particular form of methyl cellulose that
also incorporates 8% of hydroxypropyl methylcellulose.)

**Polyethylene glycol 500** is a long chain polymer that forms the
hardened surface of the tablet.

**Mastercote® pink FA0430** is a mixture of white titanium dioxide and deep red erythrosine, blended to give the pink
colour of the tablets. Mastercote® is the trade name.

There are other brands of ibuprofen such as Nurofen® and
pharmacist and supermarket own-brands.

## Disprin®

There are forms of aspirin that can be dissolved in water
and taken that way, and there is this form that comes as a
chewable tablet. Those who take aspirin everyday as a pre-
ventative should take the 75 mg dose, often referred to as junior
aspirin, rather than the 300 mg dose, which is the painkiller
version described here.

*Contents*

Aspirin 300 mg            Sodium lauryl sulphate
Calcium carbonate         Saccharin
Maize starch              Crospovidone
Citric acid               Lime flavour
Talc

**Aspirin** is the active ingredient and there are 300 mg of this.
Aspirin blocks both COX1 and COX2 enzymes, which is why
it causes bleeding.

**Calcium carbonate** bulks out the tablet.

**Maize starch** is an edible filler.

**Citric acid** gives the tablet a pleasant flavour and is neutralised
by the calcium carbonate.

**Talc** is a filler that gives the tablet a smooth feel in the mouth
when chewed.

**Sodium lauryl sulphate** is a surfactant that assists in the break-
up of the tablet.

**Saccharin** gives the tablet a sweet taste.

**Crospovidone** is a version of povidone (also known as PVP), which is a water soluble polymer that holds the tablet contents together.

**Lime flavour** consists of citral, which tastes like lemon; limonene, which is the taste of oranges; β-pinene, which smells of pine; and fenchone, which smells like camphor. Together these produce the flavour we associate with lime juice.

There are other brands of aspirin, such as Anacin®, pharmacist and supermarket own-versions, and some kinds of Alka-Seltzer® contain it.

## 2.   INDIGESTION

*The Chemistry*

Indigestion is usually caused by there being too much acid in the stomach, and this can be due in part to the things we eat and drink, and also due to excess acid being produced by the stomach itself. The easy way to treat acid indigestion is to neutralise it, and the traditional way of doing this quickly was with bicarbonate of soda (sodium hydrogen carbonate). This reacts with the acid and releases carbon dioxide, which is then emitted as a relieving burp of gas and a sign that the treatment has worked.

A more convenient remedy is to chew a carbonate salt slowly and allow the saliva to carry the relieving chemical to the stomach. The most popular type of indigestion tablet contains both calcium carbonate and magnesium carbonate.

## Rennie® Spearmint

*Contents*

| | |
|---|---|
| Calcium carbonate | Talc |
| Heavy magnesium carbonate | Povidone |
| Sucrose | Sodium saccharin |
| Glucose | Magnesium stearate |
| Spearmint flavour | |

© Shutterstock

**Calcium carbonate** is the main active ingredient and there are 680 mg of this in a 1.5 g tablet.

**Heavy magnesium carbonate** is a mixture of magnesium carbonate and magnesium hydroxide, both of which are good at neutralising acid.

**Sucrose** (sugar) is added as a bulking agent, dispersant, and sweetener. A tablet contains 0.25 g. (There are also sugar-free Rennie® tablets.)

**Glucose** is also a bulking agent and again there is 0.25 g per tablet.

**Spearmint flavour** is just what it says.

**Talc** is an excipient, added to give the dissolving tablet a smooth feel in the mouth.

**Povidone** is a water-soluble polymer that holds the other ingredients together so that the tablet does not disintegrate on storage.

**Sodium saccharin** makes the tablet taste sweet.

**Magnesium stearate** both helps the tablet to disintegrate when it is moistened by saliva and it prevents the tablets from sticking to the machinery used in their making.

## Alka-Seltzer®

This is a popular drink to relieve a hangover and the headache and stomach discomfort linked to it. It delivers both aspirin, to treat the former, and sodium hydrogen carbonate for the latter.

*Contents*

Aspirin (324 mg)
Sodium hydrogen carbonate (1625 mg)
Citric acid (965 mg)

**Aspirin** is the active ingredient.

**Sodium hydrogen carbonate** and **citric acid** react together to generate bubbles of carbon dioxide. The **citric acid** also provides a fruity flavour.

## Zantac®

*The Chemistry*

Heartburn occurs when surplus acid from the stomach is re-gurgitated into the gullet (the oesophagus) causing a painful sensation in the chest. It is possible to reduce the production of acid, and this product does just that, thanks to the ranitidine it contains.

*Contents*

Ranitidine hydrochloride (150 mg)      Methyl hydroxypropyl cellulose (E464)
Microcrystalline cellulose             Titanium dioxide E171
Magnesium stearate                     Triacetin

**Ranitidine hydrochloride (150 mg)** is the active ingredient.

**Microcrystalline cellulose** acts as an excipient and helps the tablet to break up when it reaches the stomach.

**Magnesium stearate** ensures the tablets do not stick to the machinery stamping them out.

**Methyl hydroxypropyl cellulose (E464)** acts as the coating for the tablets and yet it easily dissolves when taken.

**Titanium dioxide E171** is a white pigment.

**Triacetin** is a humectant and lubricant.

There are other brands of heartburn relief based on ranitidine, including pharmacist and supermarket own-brands.

### 3. EYE WASH

## Optrex®

*The Chemistry*

The eyes are protected by tears that have their own powerful antimicrobial agent, the enzyme lysozyme. However, this might not be enough to prevent them becoming irritated or tired when affected by some external cause. They then need to be cleaned with something that soothes. The solution used will also partly drain away through the tear duct at the corner of the eye and into the nasal passage so it has to be perfectly safe.

*Contents*

| | |
|---|---|
| Purified water | Witch hazel |
| Alcohol | Sodium borate |
| Boric acid | Benzalkonium |
| Glycerin | chloride |

**Purified water** is the solvent.
**Alcohol** serves both as a solvent for the ingredients and partly acts as a cleansing agent.
**Boric acid** is a mild antiseptic.

© Shutterstock

**Glycerin** helps to solubilise the molecules in the witch hazel extract.

**Witch hazel** (*Hamamelis*) contains several molecules but the most important ones are the polyphenols known as tannins. These are effective anti-inflammatory agents, as well as having antiviral and antibacterial powers.

**Sodium borate** is the sodium salt of boric acid. Although with no proven efficacy, this has traditionally been used as an antiseptic.

**Benzalkonium chloride** is part surfactant and part preservative.

## 4. COUGH MEDICINE

*The Chemistry*

There are two kinds of cough medicine, one that promotes discharge of phlegm from the lungs, the other to suppress an unproductive cough caused by an irritating tickle at the back of the throat. Venos® is a well-known version of the former, Tixylix® of the latter and this can also be given to children.

## Venos® Chesty Cough Expectorant

*Contents*

| | |
|---|---|
| Guaifenesin | Treacle |
| Liquid glucose | Sodium metabisulphite |

**Guaifenesin** is the active ingredient, and a typical 5 ml dose provides 100 mg.

**Liquid glucose** is the solvent and accounts for most of the syrup.

**Treacle** acts to sweeten the cough mixture to make it more palatable.

**Sodium metabisulphite** acts as a preservative.

Another popular expectorant cough mixture that contains guaifenesin is Benylin®.

## Tixylix Dry Cough Linctus®

This product can be bought over the counter as pholcodine linctus and it will suppress a tickly cough. It is also available in a form suitable for children.

*Contents*

Pholcodine
Citric acid
Glycerol
Potassium acesulfame

Maltitol
Hydroxyethyl cellulose
Sodium benzoate

**Pholcodine** is the active ingredient and while it is based on morphine it has neither the painkilling nor the euphoric effects of that opiate. It targets the coughing reflex. It has a bitter taste, hence the need to add the artificial sweetener potassium acesulfame.

**Citric acid** has a pleasing fruity acid tang and is naturally present in lemons and limes.

**Glycerol** dissolves the pholcodine, which is not very soluble in water.

**Potassium acesulfame** is an artificial sweetener that is soluble in glycerol and unaffected by acids, and ideal for products with a long shelf-life.

**Maltitol** has laxative properties whereas pholcodine can be constipating. Although it is a sweet-tasting sugar-alcohol molecule, maltitol does not cause tooth decay.

**Hydroxyethyl cellulose** is a thickening agent that also acts as a laxative.

**Sodium benzoate** acts as a preservative and is an effective antibacterial and antifungal agent. It keeps the linctus free of microbial contamination.

## 5. EAR WAX REMOVER

**Otex**®

*The Chemistry*

Ear wax slowly oozes from sebaceous glands in the ear and is a mixture of squalene, lanosterol, and cholesterol. Depending on its composition, earwax can vary from a runny oil to a hard plug, even blocking the auditory canal. This product dissolves hard wax and disinfects the ear. (The traditional way of removing the wax is with warm olive oil and indeed this works well because it dissolves the wax.)

Ear wax remover needs to contain a safe solvent that will dissolve wax and an antimicrobial agent to keep it germ-free, which is just what the following remedy contains.

*Contents*

Urea-hydrogen peroxide, UHP
Hydrogen peroxide, $H_2O_2$

8-Hydroxyquinoline
Glycerol

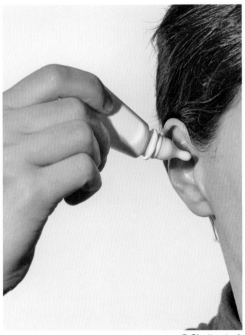

© Shutterstock

**Urea-hydrogen peroxide, UHP,** of which there is 5%, is a 1 : 1 combination of urea and hydrogen peroxide. In the ear it penetrates the earwax and there it reacts to release bubbles of oxygen gas that break up the wax. It is for this reason that the ear must not be plugged as the gas will then build up a pressure that could damage the ear drum. There is also some free **hydrogen peroxide** in the product.

**8-Hydroxyquinoline** is an antimicrobial agent, which also acts as a preservative.

**Glycerol** is the solvent that softens and breaks up the wax.

## 6. CRACKED SOLES OF FEET

*The Chemistry*

Walking can be painful when deep cracks appear on the thickened part of the skin that forms the heel. What is needed is something to prevent the skin drying out and enable it to heal itself.

### Flexitol® Heel Balm

*Contents*

| | |
|---|---|
| Aqua | Dimethicone |
| Urea | Carbomer |
| Lanolin | Triethanolamine |
| Petrolatum | Aluminium stearate |
| Decyl oleate | Methyl, ethyl, butyl, and isobutyl |
| Glyceryl oleate | parabens |
| Dicocoyl pentaerythrityl distearyl | Propyl gallate |
| citrate | Citric acid |
| Propylene glycol | BHA |
| Cera microcristallina | Parfum |

**Aqua** is water and acts as a solvent.

**Urea** helps to rehydrate dry skin.

**Lanolin** is excreted by the hair follicles of sheep and it makes their wool waterproof. It is easily absorbed by the cracked skin.

**Petrolatum** is also known as petroleum jelly (one popular brand is Vaseline®) and is a mixture of medium chain hydrocarbons extracted from crude oil and purified.

**Decyl oleate** and **glyceryl oleate** make the skin feel smooth.

**Dicocoyl pentaerythrityl distearyl citrate** is used to soften the skin.

**Propylene glycol** stops the skin from drying out.

**Cera microcristallina** is a long chain hydrocarbon wax from oil, and its crystals are much smaller than those of normal hydrocarbon waxes.

**Dimethicone** is silicone oil and it locks moisture within the skin's outer layer.

**Carbomer** is a polymer added to make the product form a smooth gel.

**Triethanolamine** is used to solubilise the petrolatum and make it mix with water.

**Aluminium stearate** is a thickening agent.

**Methyl, ethyl, butyl, and isobutyl parabens** are there as preservatives for the various ingredients, and they prevent microbes from breeding in the balm.

**Propyl gallate** and **BHA** are antioxidants.

**Citric acid** keeps the skin slightly acidic.

## 7. RHEUMATISM: AVOIDANCE AND RELIEF

*The Chemistry*

Rheumatism registers as pain at various joints of the body where the cartilage has started to wear thin. Cartilage acts as a cushion between bones that move. (Arthritis is the extreme situation when bone is rubbing against bone and this needs medical treatment.) There are remedies that offer long-term relief from rheumatism, or may even prevent its ever becoming a serious problem, and there are treatments that deal with a particular painful joint that has been affected. The first of these is a treatment that may help prevent the breakdown of cartilage by providing an essential chemical, glucosine, that it needs. The second offers immediate relief where it is most needed.

## Boots® Glucosamine Sulphate

*Contents*

Glucosamine sulphate dipotassium chloride (shellfish)
Cellulose
Polyvinylpyrrolidone
Ascorbic acid
Hydroxypropyl methyl cellulose
Glucose
Magnesium stearate
Stearic acid
Talc

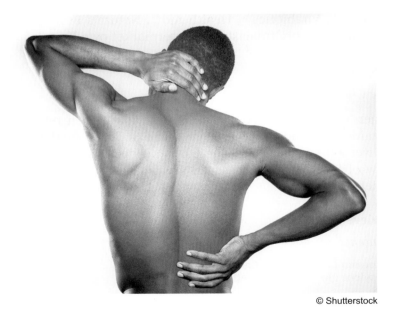

© Shutterstock

**Glucosamine sulphate dipotassium chloride (shellfish)** is the essential cartilage chemical and is extracted from shrimp shells by means of sulfuric acid and potassium chloride.

**Cellulose** helps the tablet disintegrate when it reaches the stomach.

**Polyvinylpyrrolidone** is a water-soluble polymer that holds the other ingredients together so that the tablet does not dry out and disintegrate on storage.

**Ascorbic acid** is vitamin C and it is involved in many functions such as protecting the joints, as well as being essential for tissue repair.

**Hydroxypropyl methyl cellulose** acts as a controlled release agent.

**Glucose** and **talc**, which is magnesium silicate, formula $Mg_3Si_4O_{10}(OH)_2$, are fillers.

**Magnesium stearate** and **stearic acid** prevent the tablets from sticking to the equipment that compresses them into tablet form.

There is also a similar product that is a combination of glucosamine and chondroitin. The latter is also a component of cartilage.

Glucosamine sulphate is also sold under the trade names of Jointace®, Vitabiotics®, Health Aid®, Nature's Best®, *etc.*, and there are also pharmacist and supermarket own-brands.

## Rheumatic Rubs

*The Chemistry*

When pain is localised to a muscle or joint, then relief can be obtained by rubbing it with a gel that can penetrate the skin and deliver a pain reliever to where it is needed. The increase in the flow of blood to the massaged area may also offer the desired relief.

## Ibuleve™ Gel

*Contents*

| | |
|---|---|
| Ibuprofen (5.0%) | Carbomer |
| Propylene glycol | Diethylamine |
| IMS | Purified water |

**Ibuprofen (5.0%)** is the pain reliever.

**Propylene glycol** acts as a solvent for the ibuprofen.

**IMS** is industrial methylate spirit, which is ethanol (alcohol).

**Carbomer** is the polymer polyacrylate and it is added to make the product into a smooth gel.

**Diethylamine** is used to assist the ibuprofen in penetrating the skin.

## Voltarol®

This is a rubbing gel that can be applied externally to the site of the pain. It is also available in tablet form.

*Contents*

| | |
|---|---|
| Diclofenac diethylammonium | Liquid paraffin |
| Carbomer | Perfume |
| Macrogol cetostearyl ether | Propylene glycol |
| Cocoyl caprylocaprate | Purified water |
| Isopropyl alcohol | |

**Diclofenac diethylammonium** is a non-steroid anti-inflammatory with the diclofenac being the active component. This blocks the cyclo-oxygenase enzyme COX2 that causes the pain but not the COX1 enzyme that protects the gut.

**Carbomer** is added as a thickening agent and emulsifier to help the other ingredients to blend smoothly together.

**Macrogol cetostearyl ether** consists of polyethylene glycol with fatty alcohols attached. It makes an ideal base for ointments because it is soluble in water and other solvents.

**Cocoyl caprylocaprate** acts to blend all the ingredients together.

**Isopropyl alcohol** is a solvent and helps the gel to be absorbed.

**Liquid paraffin** acts as a solvent for some of the ingredients. It acts as a lubricant when applying the cream to the skin.

**Propylene glycol** helps to keep all the other ingredients suspended in the gel.

## 8.   SKIN INFECTIONS

*The Chemistry*

The skin can become the host to a parasitic growth such as a fungus or a virus. Common invaders are fungi that cause rashes, and viruses that cause warts and verrucas. What is needed is something, such as an acid, that will kill the intruder, but will not damage living skin.

### Bazuka™ Gel

Care must be taken not to apply this directly to unaffected skin because it is quite a strong acid. The gel is dispensed through a fine nozzle and must be carefully targeted to cover just the wart or verruca.

*Contents*

| | |
|---|---|
| Salicylic acid | Pyroxylin |
| Camphor | Ethanol |
| Povidone | Acetone |

**Salicylic acid** comprises 26% of the contents. This chemical has the ability to lift off the top layer of dead skin cells and so allows it to get to work on a verruca, which it destroys by interfering with its enzymes. When applied to a wart it carries out the same function but more directly.

**Camphor** is readily absorbed through the skin and so helps the active ingredient, salicylic acid, to work more quickly.

**Povidone** is a water-soluble polymer that acts as the glue that sticks the Bazuka™ to the verruca or wart.

**Pyroxylin** makes camphor more pliable.

**Ethanol** and **acetone** act as solvents for the various ingredients, and especially for the salicylic acid. These are low boiling liquids and so evaporate quickly leaving behind a film of polymer that protects the Bazuka™ while it gets to work.

### Canesten®

This is an anti-fungus cream that will deal with athlete's foot. It will also treat ringworm and sweat rash if they are due to a fungus, as well as certain kinds of rash on the external parts of the male and female genitals.

*Contents*

| | |
|---|---|
| Clotrimazole (1%) | Cetyl palmitate |
| Benzyl alcohol | Cetostearyl alcohol |
| Polysorbate 60 | Octyldodecanol |
| Sorbitan stearate | Purified water |

**Clotrimazole (1%)** is a powerful antifungal chemical.

**Benzyl alcohol** acts partly as a solvent for other ingredients but also prevents the growth of bacteria.

**Polysorbate 60** is an emulsifier.

**Sorbitan stearate** acts as an emulsifier.

**Cetyl palmitate, cetostearyl alcohol** and **octyldodecanol** are emollients and skin conditioners, the last of which also has a pleasing aroma.

## 9.   ANTISEPTIC OINTMENT

*The Chemistry*

The skin is the largest organ of the human body and prone to damage by a cut, a graze, an insect bite, or a microorganism. Then we can help the skin heal itself more quickly by ensuring that the wound does not become infected, and the easiest way to do that is to rub it with an antiseptic ointment. The following one has been around for more than 50 years and is well known.

## Savlon®

*Contents*

| | |
|---|---|
| Cetrimide (0.5%) | Methyl hydroxybenzoate |
| Chlorhexidine gluconate (0.1%) | Propyl hydroxybenzoate |
| Cetostearyl alcohol | Fragrance |
| Liquid paraffin | Purified water |

**Cetrimide (0.5%)** is the active agent and its molecule carries a positive charge. As such, it attaches itself to the cell membrane of a microbe, which it distorts and weakens and so

enters the cell. Once inside it interferes with vital proteins that the cell needs and the cell dies.

**Chlorhexidine gluconate (0.1%)** is also a positively charged antiseptic and is especially deadly to bacteria.

**Cetostearyl alcohol** acts both as a thickener and as an emollient.

**Liquid paraffin** acts as a solvent for some of the ingredients. It acts as a lubricant when applying the cream to the skin.

**Methyl hydroxybenzoate** and **propyl hydroxybenzoate** are antibacterials.

**Fragrance** is not identified as such.

**Purified water** is water that has been made sterile.

There are other products for helping damaged skin including washing it with a disinfectant solution such as Dettol® or TCP™, or applying a protective cream such as Germolene® or Sudocrem®.

## 10.   DIARRHOEA

*The Chemistry*

The too rapid passage of the contents of the stomach through the intestine may be due to several things, some causes of which need medical attention. However, we can all experience this unpleasantness occasionally and it is generally due to inflammation caused by microbial infections. The condition generally clears up within a few days and this time may be shortened by taking the appropriate medication. Persistent diarrhoea, however, may hide symptoms of a serious disease and sufferers should seek medical advice.

### Imodium®

There are two versions of this, one that clears up diarrhoea and the other that deals also with a bloated stomach caused by retention of gas. It is the latter type that we consider here.

*Contents*

| | |
|---|---|
| Loperamide HCl (2 mg) | Dibasic calcium phosphate |
| Simethicone (125 mg) | Flavour |
| Acesulfame potassium | Microcrystalline cellulose |
| Croscarmellose sodium | Stearic acid |

**Loperamide HCl (2 mg)** is the active agent and this acts by blocking certain receptors of the intestine wall and thereby deactivates them so that they don't release the prostaglandins that are responsible for its over-activity.

**Simethicone (125 mg)** is silicone oil and this acts as a defoaming agent that collapses the bubbles of gas in the stomach and so ejects them as wind.

**Acesulfame potassium** is an artificial sweetener.

**Croscarmellose sodium** is a form of cellulose in which the chains are cross-linked. This causes the tablet to swell when it becomes moist in the stomach and so helps release its contents.

**Dibasic calcium phosphate** acts as filler.

**Flavour** is unspecified.

**Microcrystalline cellulose** helps the tablet to disintegrate when it reaches the stomach.

**Stearic acid** prevents the capsules sticking to the machinery that makes them.

## 11.  SORE THROAT REMEDIES

*The Chemistry*

An inflamed pharynx, caused by a virus or bacteria, makes swallowing painful. It will generally cure itself in a few days and the best remedy is to drink more fluids and suck a soothing lozenge. This can be an antiseptic lozenge, which will slowly release a chemical that will help eradicate the offending microbes that are causing the irritation and soreness. Ideally it will also have a painkilling ingredient, as does the following popular kind.

### Strepsils®

*Contents*

| | |
|---|---|
| Hexylresourcinol | Blackcurrant flavour (containing propylene glycol) |
| Sucrose | Carmoisine edicol (E122) |
| Glucose | Patent blue V (E131) |
| Levomenthol | |

**Hexylresourcinol** has both antiseptic and anaesthetic properties. It is recognised as safe to consume and has the food code E586. There are 2.4 mg per tablet.

**Sucrose** and **glucose** make the tablet pleasant to suck.

**Levomenthol** occurs naturally in peppermint oil. It triggers the cold-sensitive receptors in the mouth, producing a fresh feel and may help breathing.

**Blackcurrant flavour (containing propylene glycol)** is the flavouring and the propylene glycol is a solvent for it.

**Carmoisine edicol** (food code **E122**) is a red dye.

**Patent blue V** (food code **E131**) is a blue dye.

There are several other sore throat lozenges including Merocets®, Beechams®, Tyrozets®, Halls®, and pharmacist and supermarket own-brands. These also contain a combination of antiseptic and analgesic agents.

## 12.   ANTI-ITCHING CREAM

*The Chemistry*

There are all kinds of reasons why our skin itches. It might even be caused by parasites, namely the scabies mite or the lice that infest the genital region and are known as crabs. These creatures burrow beneath the skin and are reputed to cause unbearable itching.

Normally if the skin itches, a scratch will stop it but sometimes the itching is more persistent and due to being bitten by an insect, or to inflammation caused by an infection. Then you need to spread an anti-itching cream on the affected area. Some are based on emollients and rehydrate the skin, which is often sufficient to stop the itch, some rely on an anaesthetic effect to deaden the sensation of itching, some contain a steroid, which will reduce inflammation and so relieve the itching, and some will exterminate the root cause if this is due to a parasite. The following product is of this last kind.

### Eurax® Cream

This product will reduce itching, whatever its cause, and is deadly to mites.

*Contents*

| | |
|---|---|
| Crotamiton (10%) | Stearyl alcohol |
| Methyl hydroxybenzoate | Strong ammonia solution |
| Phenylethyl alcohol | Stearic acid |
| Glycerol | Hard paraffin |
| Triethanolamine | White beeswax |
| Sodium lauryl sulfate | Perfume |
| Ethylene glycol monostearate | |

**Crotamiton (10%)** is a drug that is absorbed by the skin and so can control the inflammation or irritation that we perceive as itching. It also kills mites.

**Methyl hydroxybenzoate** is methyl paraben and is a preservative.

**Phenylethyl alcohol** has antimicrobial properties.

**Glycerol** acts as a solvent.

**Triethanolamine** acts as an emulsifier.

**Sodium lauryl sulfate** is a surfactant.

**Ethylene glycol monostearate, stearyl alcohol** and **stearic acid** are emollients, thickeners and emulsifiers.

**Strong ammonia solution** makes the cream slightly alkaline and stabilises the pH.

**Hard paraffin** consists of long chain hydrocarbons extracted from oil. It acts as a lubricant when applying the cream to the skin.

**White beeswax** is natural beeswax that has been bleached to make it white.

**Perfume** is unspecified.

Other anti-itching creams include Lanacane® (which contains the anaesthetic benzocaine), Cortizone-10® (which contains the steroid hydrocortisone), Oilatum® and Boots® Expert Anti-Itch Soother (both of which rely on emollients).

CHAPTER 2

# The Utility Room

The utility room is where the washing machine, tumble dryer, and central heating boiler are located, along with the products

Chemistry at Home: Exploring the Ingredients in Everyday Products
By John Emsley
© John Emsley, 2015
Published by the Royal Society of Chemistry, www.rsc.org

associated with their use. It is also where general cleaning agents are stored. The ones discussed in this chapter are as follows.

1. Laundry detergents (Ariel® Excel Gel, Persil® Non-Bio Washing Powder, Persil® Small & Mighty Non-Bio)
2. Fabric softener (Comfort®)
3. Tumble dryer fragrancer (Lenor® Tumble Dryer Sheets)
4. Surface cleaners (Cillit Bang® Power Cleaner Grease & Floor, Flash® Clean & Shine, Cillit Bang® Power Cleaner)
5. Stain-removers (Vanish Oxi Action® Powder, Shout® Laundry Stain-Removing Spray, Shout® Carpet Stain-Remover)
6. Window cleaner (Windolene® Trigger (Spray), Windolene® Cream, Windolene® Wipes)
7. Drain unblocker (Mr Muscle® Sink & Drain Gel)
8. Carbon monoxide detector (various kinds)

# 1.  LAUNDRY DETERGENTS

*The Chemistry*

Laundry detergents were introduced early last century, and Persil® became the leading brand. This consisted of soap powder as the surfactant and sodium perborate as the bleaching agent. It worked superbly well because clothes then were washed at high temperatures, often in almost boiling water, and this activated the percarbonate to produce the hydrogen peroxide that was the bleaching agent.

In the 1950s, chemical surfactants were introduced because, unlike soap, these did not form a scum in hard water, which is the kind of water supplied to most homes in the UK. Then, in 1973, came the Oil Crisis, which forced up the cost of energy and people began to wash clothes at much lower temperatures, a trend that has intensified over the past five years.

In the 1960s, enzymes were added to laundry products like Ariel®, and these removed stains of the biological kind. However, in the 1980s, a group of consumers claimed that the enzymes were causing rashes on their skin. Apparently this did not happen with the traditional kind of detergent so these continued to be sold and are now labelled as 'non-bio'. They still account for a sizeable fraction of the market. (Dermatologists investigated the claims of the protestors but found no scientific support for enzymes being to blame.)

Today most washing machines are operated at 40 °C or less, and while 'non-bio' detergents work better at 60 °C, there are even versions of these that can be used at 40 °C, and one of these is described below.

Modern detergents contain the following: surfactants to remove greasy dirt; water softeners such as zeolites or chelants to deactivate interfering metal ions; enzymes to digest biological stains; a bleaching agent and bleach activator; anti-redeposition agents to hold the dirt in the wash water once it's been released from fabrics; and an optical brightener. (This last type of molecule clings to fibres and absorbs invisible UV light and re-emits it as white light so that fabrics look brighter.) They will also contain an added fragrance such as alpha-isomethyl ionone, which has a sweet violet odour.

Detergents come in all shapes and sizes, ranging from so-called 'big box' powders to some that come in capsule-size doses. The following are representative.

© Shutterstock

## Ariel® Excel Gel

This laundry detergent can be used at temperatures as low as 15 °C but it does not contain a bleaching agent. The contents as listed on the packaging itself are only those needed to meet legal requirements. The complete list can be obtained from the Proctor & Gamble website and this list is as follows:

### Contents

Aqua (water)
Sodium laureth sulfate
MEA dodecylbenzenesulfonate
MEA palm kernelate
MEA citrate
C12–14 pareth-7
PEI-ethoxylate
Propylene glycol
Glycosidase
Protease

Disodium distyrylbiphenyl disulfonate
Trimonoethanolamine etidronate
MEA borate
Sodium sulfate
Colorant
Butylphenyl methylpropional
Citronellol
Eugenol
Geraniol
Linalool

**Sodium laureth sulfate**, which is made from coconut oil, **MEA dodecylbenzenesulfonate**, which comes from fossil oil, and **MEA palm kernelate**, which is made from palm oil, are anionic surfactants.

**MEA citrate** and **trimonoethanolamine etidronate** chelate the calcium and magnesium ions of hard water so they cannot interfere with the surfactants or deposit on the washing machine as limescale.

**C12–14 pareth-7** and **propylene glycol** are non-ionic surfactants and they help keep all the other ingredients suspended in the gel.

**PEI-ethoxylate** acts to prevents dirt and dye molecules that have been washed off fabrics from depositing themselves on other fabrics.

**Glycosidase** and **protease** are the enzymes that digest food stains and personal residues on underclothes. Glycosidase digests carbohydrates and protease digests proteins.

**Disodium distyrylbiphenyl disulfonate** acts as an optical brightener.

**MEA borate** is there to keep the pH of the gel slightly alkaline, to protect the enzymes.

**Sodium sulfate** is a bulking agent.

Perfume raw material is a complex mixture, some items of which have to be identified in a label because some people may be sensitive to them. The ones so identified here are: **butylphenyl methylpropional** that has a clean smell like washing dried in the open air, while **citronellol** smells of geraniums and roses, **eugenol** has a spicy smell like cloves, **geraniol** smells of roses, and **linalool** has a floral spicy smell.

Other liquid detergents of this type include Persil® and Bold®.

### Persil® Non-Bio Washing Powder

This is known in the trade as a 'big box' detergent and caters both for those who find that their skin is sensitive to biological detergents and those who traditionally like to use lots of product, in the belief that the more you use, the cleaner your wash will be.

## Contents

Sodium sulfate
Sodium carbonate
Sodium dodecylbenzene-
  sulfonate
Sodium silicate
Sodium carbonate peroxide
Zeolite
Citric acid
TAED
C12–15 pareth-7
Sodium acrylic acid/MA
  copolymer
Corn starch modified
Cellulose gum
Ceteareth-25
Parfum
Sodium chloride

Tetrasodium etidonate
Calcium sodium EDTMP
Disodium anilinomorpholinotriazinyl-
  aminostilbenzenesulfonate
Polyethylene terephthalate
Phenylpropyl ethyl methicone
Sodium bicarbonate
Bentonite
Glyceryl stearates
Sodium polyacrylate
Disodium distyrylbiphenyl
  disulfonate
Aluminium silicate
Sodium polyaryl sulfonate
Imidazolidinone
Polyoxymethylene melamine

**Sodium sulfate** ensures the powder is free-flowing.

**Sodium carbonate, sodium silicate** and **sodium bicarbonate** partly act as fillers, providing bulk, but more importantly they make the wash water alkaline and this ensures that the hydrogen peroxide released from the sodium carbonate peroxide will react with the TAED better and so boost the production of the real bleaching agent, peracetic acid.

**Sodium dodecylbenzenesulfonate** is an anionic surfactant.

**Sodium carbonate peroxide** is the bleaching agent but only behaves as such when it reacts with **TAED** (tetra-acetyl-ethylene-diamine) to form sodium peracetate, which is the active bleaching agent.

**Zeolite** is a water softener and it does this by trapping calcium ions.

**Citric acid, tetrasodium etidonate** and **calcium sodium EDTMP** chelate the calcium of the hard water.

**C12–14 pareth-7** and **ceteareth-25** are non-ionic surfactants.

**Sodium acrylic acid/MA copolymer, cellulose gum** and **polyethylene terephthalate** are anti-redeposition agents.

**Corn starch modified** ensures that there are no active enzymes in the product.

**Parfum:** the fragrance molecules are not specified so do not include those that might affect some people.

**Sodium chloride** is salt.

**Disodium anilinomorpholinotriazinylamino-stilbenzenesulfo-
nate** and **disodium distyrylbiphenyl disulfonate** act as optical brighteners. They cling to fibres and absorb invisible
UV light and re-emit it as white light so that white fabrics
look whiter.

**Phenylpropyl ethyl methicone** is a silicone oil that prevents the
surfactants from forming too much foam.

**Bentonite** is a mineral that clings to the fabric in the wash and
makes it feel softer.

**Glyceryl stearates** help keep the powder free-flowing.

**Sodium polyacrylate** attracts water and helps keep the
powder dry.

**Aluminium silicate** and **polyoxymethylene melamine** prevent
the solid ingredients in the pack from caking.

**Sodium polyaryl sulfonate** is an anionic surfactant.

**Imidazolidinone** is a solvent.


## Persil® Small & Mighty Non-Bio

This comes as a gel and is designed for washing at lower temperatures, such as 40 °C.

*Contents*

Aqua
MEA dodecylbenzenesulfonate
Sodium laureth sulfate
C11–15 pareth-7
Propylene glycol
TEA-hydrogenated cocoate
MEA-citrate
Aziridine homo-polymer
  ethoxylated
Etidronic acid
Triethanolamine
Acrylates copolymer
Parfum
Styrene/acrylates copolymer
MEA-sulfate
Sodium sulfite
Disodium distyryrlbiphenyl
  disulfonate
Sodium sulfate
Sodium lauryl sulfate

**Aqua** is water and **propylene glycol** is also a solvent.

**MEA dodecylbenzenesulfonate, sodium laureth sulfate** and
**sodium lauryl sulfate** are anionic surfactants.

**C14–15 pareth-7** and **TEA-hydrogenated cocoate** are non-ionic
surfactants.

**MEA-citrate** and **etidronic acid** act as chelating agents.

**Aziridine homo-polymer ethoxylated** is an anti-redeposition
agent.

**Triethanolamine** acts partly to solubilise other ingredients and partly to act as a non-ionic surfactant.

**Parfum:** the fragrance molecules are not specified so do not include those that might affect some people.

**Acrylates copolymer** thickens the viscosity of the detergent.

**Styrene/acrylates copolymer** makes the detergent opaque, which is how the consumer expects it to look.

**MEA-sulfate** is monoethanolamine sulfate and is a surfactant.

**Sodium sulfite** prevents microbes from contaminating the product.

**Disodium distyryrlbiphenyl disulfonate** is the optical brightener.

**Sodium sulfate** is a filler.

Other low temperature detergents are available such as Ariel®, Surf® and Daz®.

## 2.  FABRIC SOFTENERS

*The Chemistry*

Fabric conditioners are designed to make clothes and bedding feel soft and they do this by coating them with a film of molecules that have a hydrocarbon chain attached. These molecules are positively charged, or have an atom that is drawn to a negative site, and as such are attracted to the surface of fibres, which generally are negatively charged.

Fabric conditioners should not be used to make towels feel softer because their hydrocarbon film will tend to repel water, which is not what a towel should do.

## Comfort®

*Contents*

Aqua
Dihydrogenated tallowoylethyl
  hydroxyethylmonium methosulfate
Isopropyl alcohol
Cetearyl alcohol
Parfum
Laureth-20

Benzisothiazolinone
Trimethylsiloxysilicate
Hydrogenated vegetable
  glycerides
Glycol stearate
Cellulose gum

**Aqua** and **isopropyl alcohol** are solvents.

**Dihydrogenated tallowoylethyl hydroxyethylmonium methosulfate** is what forms the soft coating on fibres. It is derived from tallow, which comes from the hard fat of cows and pigs.

**Cetearyl alcohol** is a mixture of long chain hydrocarbons.

**Parfum:** the fragrance molecules are not specified so do not include those that might affect some people.

**Laureth-20** is a non-ionic surfactant with a long hydrocarbon chain attached.

**Benzisothiazolinone** is an antibacterial preservative.

**Trimethylsiloxysilicate** is an antifoaming agent.

**Hydrogenated vegetable glycerides** are oils whose unsaturated fatty acids have reacted with hydrogen thereby preventing them from going rancid.

**Glycol stearate** and **cellulose gum** are emollients.

Other brands of fabric softeners include Fairy®, Lenor®, and supermarket versions.

### 3.  TUMBLE DRYER FRAGRANCER

*The Chemistry*

When washing is dried in the open air on a sunny, breezy day it acquires a smell that seems clean, fresh, and hygienic. This is thought to be due to molecules on fabrics being oxidised by exposure to UV rays. Nevertheless, the effect is seen as pleasant and nostalgic.

Most washing is now dried indoors using a dryer and as such it does not emerge with the same fresh smell. However, fragrance chemists have developed a range of molecules whose odour simulates the smell of clothes dried outdoors. These can be impregnated on to a sheet of material that is put in the dryer with the washing and will transfer its fragrance molecules to the clothes, towels, or bedding being dried. They also make ironing more fragrant.

## Lenor® Tumble Dryer Sheets

*Contents*

Surfactant                     Amyl cinnamal
Perfumes                       Benzyl benzoate
Alpha-isomethyl ionone         Hexyl salicylate

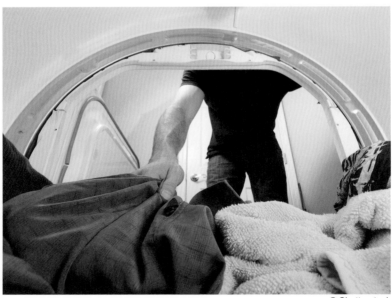

© Shutterstock

Butylphenyl methylpropional          Hydroxyisohexyl 3-cyclohexene
Citronellol                                            carboxaldehyde
Coumarin                                           Limonene
Eugenol                                             Linalool
Hexyl cinnamal

**Surfactant** is not specified but is assumed to assist the fragrance molecule to cling to the sheets.

**Perfumes** describes a complex mixture, some items of which have to be identified in a label because some people may be sensitive to them. The ones identified here are: **alpha-isomethyl ionone** and **butylphenyl methylpropional,** which together have a clean smell like washing dried in the open air; **amyl cinnamal,** which smells like jasmine; **citronellol,** which smells of geraniums and roses; **coumarin,** which has the scent of new-mown hay, **eugenol,** which has a spicy smell like cloves; **hexyl cinnamal,** which smells of camomile; **hydroxyisohexyl 3-cyclohexene carboxaldehyde**, which smells of lily of the valley; **limonene**, which smells of oranges; and **linalool**, which has a floral spicy smell.

**Benzyl benzoate** and **hexyl salicylate** act as preservatives.

## 4.   SURFACE CLEANERS

*The Chemistry*

The grime that we encounter on surfaces in our home comes as dirt on floors, food residues on kitchen surfaces and counters, and grease on hobs and handles. To remove it, we need a surfactant to solubilise the grease that binds it to the surface. For most cleaners it is better to have an alkaline solution. We also want the product to smell nice.

There are dozens of surface cleaners on the market. Here we look at three, and there are many alternatives of the same brand, each offering some extra benefit.

### Cillit Bang® Power Cleaner Grease & Floor – Citrus Force

*Contents*

Aqua
C12–C18 alkyldimethyl amine *N*-oxides
Monoethanolamine 2-aminoethanol
Water soluble polycarboxylate
Sodium bicarbonate
Sodium C10–C14 alkyl benzenesulfonate

Parfum
Limonene
Sodium hydroxide
Colorant
Antifoam silicone emulsion.

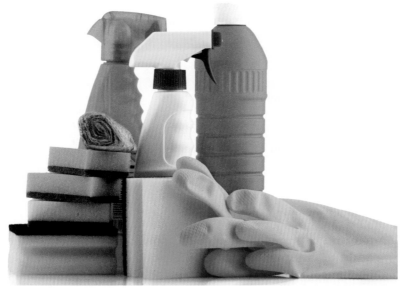

© Shutterstock

**Aqua** is water, the main solvent.

**C12–C18 alkyldimethyl amine *N*-oxides** are a powerful group of non-ionic surfactants.

**Monoethanolamine,** also known as 2-aminoethanol, is used to keep the liquid alkaline and help solubilise grease.

**Water soluble polycarboxylate** acts as a thickening agent.

**Sodium bicarbonate** and **sodium hydroxide** keep the liquid alkaline.

**Sodium C10–C14 alkyl benzenesulfonate** is an anionic surfactant.

**Parfum** is a complex mixture, some items of which have to be identified in a label because some people may be sensitive to them. The one so identified here is **limonene**, which smells of oranges.

**Colorant** is not specified.

**Antifoam silicone emulsion** is added so that the surfactants do not produce foam, which would reduce the amount of cleaner in contact with the surface.


### Flash® Clean & Shine

The components listed below are taken from the Proctor & Gamble website, which you access *via* an eight digit number on the product itself, in this case 96257325. Six of its ingredients have to be identified as fragrances.

*Contents*

| | |
|---|---|
| Aqua | Citral |
| C9–11 Pareth-n | Hexyl cinnamal |
| Sodium dodecylbenzenesulfonate | Linalool |
| Lauramine oxide | Limonene |
| Sodium palm kernelate | Citronellol |
| Parfum | Dipropylene glycol |
| Methyacrylic/acrylic acid copolymer modified | Geraniol |
| Carbonate salt | Glutaral |
| Sodium citrate | Benzisothiazolinone |
| Sodium diethylenetriamine pentamethylene | Colorant |
|   phosphonate | Colorant |
| Butyloctanol | |

**Aqua** and **dipropylene glycol** are solvents for the ingredients and the latter may also help remove greasy marks.

**C9–11 Pareth-n** is a non-ionic surfactant.

**Sodium dodecylbenzenesulfonate** is an anionic surfactant.

**Lauramine oxide** is a non-ionic surfactant.

**Sodium palm kernelate** is an anionic surfactant derived from plant oils.

**Parfum** is a complex mixture, some items of which have to be identified in a label because some people may be sensitive to them. The ones identified here are **citral**, which smells of lemon; **hexyl cinnamal,** which smells of camomile; **linalool,** which has a floral spicy smell; **limonene,** which smells of oranges; **citronellol,** which smells of geraniums and roses; and **geraniol,** which smells of roses.

**Methyl acrylic/acrylic acid copolymer modified** acts as a water softener and is also a thickening agent.

**Carbonate salt** is sodium carbonate, which has some cleaning ability.

**Sodium citrate** acts both as an emulsifier and keeps the pH stable.

**Sodium diethylenetriamine pentamethylene phosphonate** is good at cleaning glass.

**Butyloctanol** is the solvent for the various fragrances.

**Glutaral** is a disinfectant.

**Benzisothiazolinone** is an antibacterial agent.

**Colorant.** Two are indicated in the ingredient list but neither is identified.


## Cillit Bang® Power Cleaner

Sometimes limescale is a problem. Then we need an acid-containing cleaner that can react chemically with it and thereby dissolve it.

*Contents*

| | |
|---|---|
| Aqua (water) | C9–11 pareth-6 |
| Sulfamic acid | Parfum |
| Oxalic acid | Hexyl cinnamal |

**Aqua** (water) acts as a solvent.

**Sulfamic acid** is good at cleaning metals and, because it is an acid, it dissolves limescale.

**Oxalic acid** acts both as an acid and as a sequestrant for calcium ions, which interfere with the surfactant.

**C9–11 pareth-6** is a surfactant.

**Parfum.** Perfume raw material is a complex mixture, some items of which have to be identified in a label because some people may be sensitive to them. The one so identified here is **hexyl cinnamal,** which smells of camomile.

## 5. STAIN-REMOVERS

*The Chemistry*

Fabrics and carpets are easily stained and there are products designed to remove or decolorize the offending molecules. Sodium percarbonate removes bleachable stains because it reacts with another ingredient, TAED, and forms peracetic acid, and it is this chemical that will remove those stains that are highly coloured. Unlike household bleach, which relies on the powerful oxidising agent sodium hypochlorite, peracetic acid is not as strong as bleach so less likely to damage fabric dyes, but it will eventually cause fading of bright dyes. (For this reason some so-called 'colour' detergents are made without bleaches especially for washing such items of clothing.)

## Vanish Oxi Action® Powder

This powder can be dissolved in water and used as such or it can be added to the washing machine along with the usual detergent to boost the amount of oxidising agent.

© Shutterstock

*Contents*

Sodium percarbonate                                  Pareth-5
Sodium carbonate                                     Zeolite
Sodium sulfate                                       Protease
Tetaacetylethylenediamine (TAED)                     Parfum
Disodium disilicate                                  Lipase
Sodium dodecylbenzenesulfonate                       Amylase

**Sodium percarbonate** is the source of peroxide.

**Sodium carbonate** makes the wash water alkaline and this
  ensures that the hydrogen peroxide released from the percarbonate will react with the TAED better and so boost the
  production of the real bleaching agent, which is peracetic acid.

**Sodium sulfate** ensures the powder is free-flowing.

**Tetraacetylethylenediamine** (TAED) reacts with the sodium
  percarbonate to form peracetic acid, which acts as the bleach.

**Disodium disilicate** protects exposed metal surfaces.

**Sodium dodecylbenzenesulfonate** is an anionic oil-based
  surfactant while **pareth-5** is a non-ionic surfactant.

**Zeolite** is a type of aluminium silicate with large cavities that can
  attract and hold calcium ions, thereby softening the water.

**Protease** will digest proteins.

Although **parfum** is indicated in the list of components, its
  actual components are not listed so do not constitute a
  threat to sensitive individuals.

**Lipase** will digest fats.

**Amylase** will digest carbohydrates.


## Shout® Laundry Stain-Removing Spray

This is a liquid designed to be sprayed directly on to stains. It
does not contain an oxidising agent so it is no threat to fabric
dyes. It relies mainly on a non-ionic surfactant to detach stains
from surfaces and an enzyme to digest them.

*Contents*

Water                          Polyacrylate polymer
Ethoxylated alcohol            Benzisothiazolinone
Citric acid                    Subtitlisin
Sodium hydroxide               Fragrance
Sodium borate                  Dodecanol
Acrylic copolymer

**Water** is the solvent.

**Ethoxylated alcohol** is a non-ionic surfactant.

**Citric acid** and **sodium hydroxide** are listed as separate chemicals but these will react chemically to form sodium citrate, which will act to remove the effects of interfering metals.

**Sodium borate** is also known as borax and is reputed to have deodorising properties.

**Acrylic copolymer** and **polyacrylate polymer** are thickeners.

**Benzisothiazolinone** keeps the stain-remover free of germs.

**Subtitlisin** is a protein-digesting enzyme.

Although **fragrance** is indicated in the list of components, its actual components are not listed so do not constitute a threat to sensitive individuals.

**Dodecanol** is an alcohol with a long hydrocarbon chain and is made from palm or coconut oil. It acts as a stabiliser.

Other stain removers include Ariel® Spray and Vanish® Spray.

## Shout® Carpet Stain-Remover

This is designed for topical application on carpets. It relies on surfactants to attack the grease, which is what generally causes unwanted grime to stick to fibres.

*Contents*

| | |
|---|---|
| Water | Ethoxylated alcohol |
| Isopropylamine dodecylbenzene sulfonate | Fragrance |
| Sodium citrate | Acrylic copolymer |
| Sodium capryl sulfonate | Methylisothiazolinone |

**Water** is the solvent.

**Isopropylamine dodecylbenzene sulfonate** and **sodium capryl sulfonate** are anionic surfactants.

**Sodium citrate** acts both as an emulsifier and as an acidity regulator keeping the pH stable.

**Ethoxylated alcohol** is a non-ionic surfactant that does not foam easily.

Although **fragrance** is indicated in the list of components, its actual components are not listed so do not constitute a threat to sensitive individuals.

**Acrylic copolymer** is both a thickener and it leaves behind a protective film on the carpet.

**Methylisothiazolinone** is an antibacterial agent.

Other carpet stain removers include Dr Beckmann® and RugDoctor®.

## 6. WINDOW CLEANERS

*The Chemistry*

In the days of coal fires and smoky towns, windows needed to be cleaned quite often, even weekly. The traditional way to do it was to use a mixture of paraffin and vinegar, and this combination of a solvent and an acid did remove the dirt. Today the atmosphere is cleaner. A layer of traffic fumes and dust may still deposit on windows but some of these windows are now self-cleaning thanks to a layer of titanium dioxide on their surface. However, this chore has still to be done on most windows, and there are products available for this purpose. These rely on solvents and surfactants to dissolve the grease and grime.

## Windolene® Trigger (Spray)

*Contents*

| | |
|---|---|
| Aqua | Acetic acid |
| Butoxypropanol | Sodium laureth sulfate |
| Methoxyisopropanol | Sodium lauryl sulfate |
| Propylene glycol | Colorant |
| Alkyl polyglucoside | |

**Aqua** is water, and a solvent.

**Butoxypropanol, methoxyisopropanol**, and **propylene glycol** act as solvents for greasy residues.

**Alkyl polyglucoside** is a neutral surfactant.

**Acetic acid** provides the acid and is what vinegar contains.

**Sodium laureth sulfate** and **sodium lauryl sulfate** are negative surfactants.

**Colorant** is not specified.

## Windolene® Cream

*Contents*

| | |
|---|---|
| Aqua | C9–11 pareth-8 |
| Naphtha (petroleum), hydrotreated heavy | Parfum |
| Kaolin | Benzisothiazolinone |
| C8–10 alkane/cycloalkane/aromatic hydrocarbons | Hydrochloric acid |
| C9–11 pareth-6 | Colorant |

**Aqua** is water and the solvent.

**Naphtha (petroleum), hydrotreated heavy** is a hydrocarbon solvent that has been reacted with hydrogen to remove unsaturated bonds, while **C8–10 alkane/cycloalkane/aromatic hydrocarbons** are similar but contain benzene type molecules whose grease dissolving properties are particularly good.

**Kaolin** is the mineral known as china clay (aluminium silicate) finely ground. It acts as a non-scratching abrasive.

**C9–11 pareth-6** and **C9–11 pareth-8** are surfactants.

Although **parfum** is indicated in the list of components, its actual components are not listed so do not constitute a threat to sensitive individuals.

**Benzisothiazolinone** is a preservative.

**Hydrochloric acid** is the common acid HCl.

**Colorant** is not specified.


## Windolene® Wipes

These offer a quick and easy, all-in-one, way of cleaning windows.

*Contents*

Aqua
Isopropyl alcohol
Mono PPG-n-butylether
Monopropylenglycol methyl ether

Sodium lauryl sulfate
Alcohols C12–14 ethyoxylated
Acetic acid

**Aqua** is water and the solvent.

**Isopropyl alcohol** and **mono PPG-n-butylether** (PPG is short for propylene glycol) are organic solvents with roughly the same dissolving properties as ethanol.

**Monopropylenglycol methyl ether** is similar.

**Sodium lauryl sulfate** is an anionic surfactant, while **alcohols C12–14 ethoxylated** are non-ionic surfactants.

**Acetic acid** provides the acid and is what vinegar contains.


There are other window cleaning products, such as those marketed under the names of Astonish®, Mr Muscle®, and Jeyes®.

## 7.   DRAIN UNBLOCKER

*The Chemistry*

When a sink plunger fails to unblock a drain, then you need to attack the blockage in a different way. Drains become blocked with things like hair and food residues, which become breeding grounds for slime moulds and these then act rather like a glue to which grease can stick. The blockage needs to be removed by attacking it chemically with a powerful antibacterial agent and strong alkali.

### Mr Muscle® Sink & Drain Gel

*Contents*

| | |
|---|---|
| Aqua | Sodium silicate |
| Sodium hypochlorite | Sodium laurate |
| Sodium hydroxide | Poly(sodium styrenesulfonate) |
| Alkyl dimethyl amine *N*-oxide | Triethyl phosphate |
| Decanoic acid, sodium salt | |

**Aqua** is water and a solvent.

**Sodium hypochlorite** is powerful enough to kill all microbes and especially the slime moulds in the blockage. It will also denature and break up hair if that is part of the blockage.

**Sodium hydroxide** is a powerful alkali that removes congealed fats by reacting with them chemically to produce water-soluble glycerol and fatty acid salts that can then be flushed away.

**Alkyl dimethyl amine *N*-oxide** is a non-ionic surfactant.

**Decanoic acid, sodium salt**, and **sodium laurate** are anionic surfactants.

**Sodium silicate** acts to make the drain unblocker gel-like. It also prevents corrosion of metal components in the sink and drain.

**Poly(sodium styrenesulfonate)** is an ionic polymeric material that is soluble in water and also helps forms the gel. It will also remove calcium, which may be present as limescale.

**Triethyl phosphate** is a powerful solvent for all kinds of materials and as such it will dissolve various components of the blockage, such as pieces of plastic.

Other brands of drain unblocker include Domestos™, Buster®, Cillit Bang®, and Ecozone®.

## 8. CARBON MONOXIDE DETECTOR

*The Chemistry*

Accidental deaths from carbon monoxide (CO) come from badly ventilated gas fires and boilers. This happens when these lack enough oxygen for complete combustion of the methane fuel, which would normally form non-toxic $CO_2$ as it burns, then starts to produce CO, which is very different. You cannot smell this gas, so it may well be present at life-threatening levels without you being aware. Wherever your boiler is located, you need to have a CO detector nearby. Such detectors can measure either a rapid build-up of the gas or a slow build-up over a period of time. Ideally you should install one that has an alarm that is set to go off when the level of CO in air reaches 70 ppm, or exceeds 30 ppm for a period of 30 days.

There are three kinds of detector. The cheapest is a visual detector and this consists of silica gel, which has silicon dioxide as the support material for either palladium sulfate or ammonium molybdate. These compounds react readily with CO to form $CO_2$ and thereby are reduced to the metal itself, palladium or molybdenum, so the detector turns from a light colour to black. These detectors only last a few months, nor do they sound an alarm, but they might confirm that there is CO present at potentially dangerous levels in a room where there is a boiler. When the CO level falls, then the detecting surface regenerates itself and the colour returns to its normal light shade.

Much better are electronic CO detectors that can sound an alarm when they detect the gas. Some can be plugged into a wall socket, others work on batteries. Some rely on detecting a colour change of the kind just described above. They do this using a light-emitting diode (LED) whose beam of light impinges on the chemical composition and is then deflected on to a photocell. If the amount of light is reduced by the darkening of the chemical, then the photocell registers the change and triggers a circuit that then sounds an alarm.

Other detectors rely on changes in electrical conductivity of a wire coated with tin oxide ($SnO_2$), which is very sensitive to any CO molecules that attach themselves to its surface. This changes the conductivity of the wire and an alarm goes off.

If you are alerted to CO by any of these detectors you should immediately open doors and windows, turn off the gas appliance, vacate the room, and call in expert help.

# CHAPTER 3

# The Bathroom

© Shutterstock

This is the room where we clean ourselves and prepare to face the day and what it might bring. We need to feel fresh and fragrant, or at least not smelly.

Fragrance molecules feature in lots of the products described in this book but they are particularly important for those in this

Chemistry at Home: Exploring the Ingredients in Everyday Products
By John Emsley
© John Emsley, 2015
Published by the Royal Society of Chemistry, www.rsc.org

chapter. Fragrances were once extracted from plants and referred to as essences or essential oils. Nowadays they are more likely to have been made by fragrance chemists and synthesised in a chemical plant. Nor need they be copies of the natural fragrances but may be molecules with more subtle aromas. Some still relate to natural scents such as limonene, which smells of oranges, but even this can be modified to produce other fragrances. Perfume raw material is a complex mixture, some items of which have to be identified on a label because there are people who may be sensitive to them.

The products we might expect to find in the average bathroom (with the brands examined) are as follows:

1. Shower gels (Dove® Deeply Nourishing Body Wash, Imperial Leather® Limited Edition Shower Crème, Rituals® Zensation Foaming Shower Gel, Radox® for Men, Radox® for Men Mint and Tea Tree Shower Gel)
2. Bath products (Johnson's® Baby Bath, Radian® B Mineral Bath)
3. Shampoo and conditioner (Pantene Pro-V®, Head and Shoulders®, L'Oréal® Elvive Nutri-Gloss® Conditioner)
4. Toothpaste (Colgate Sensitive Pro-Relief® + Whitening, Sensodyne® Original)
5. Denture cleaner (Steradent® Active Plus)
6. Mouthwash (Listerine® Original, Corsodyl® Medicated Mint Mouthwash)
7. Hair dyes (Clairol® Nice'n'Easy™ Permanent Natural Medium Mahogany Brown, Just For Men™ Black Hair Colour)
8. Shaving (Gillette® Shaving Gel, Gillette® Sensitive Balm)
9. Hair removers (Veet®, Nads® For Men Body Waxing Strips)
10. Fake tan (St Tropez® Self Tan Bronzing Lotion)
11. Sunscreen (Nivea Sun® Protect & Refresh Lotion SPF20)
12. Antiperspirant (Dove® Deodorant, Sure® Cool Blue Fast Drying)
13. Shower cleaners (Mr Muscle® Shower Shine, Scale Away®)

## 1. SHOWER GELS

*The Chemistry*

There are lots of shower products designed to clean the body, and most of them contain very similar ingredients, of which the most important are gentle surfactants, which do the cleaning. We also want the surfactant to generate foam and some modern ones are particularly good at this. However, surfactants will not only remove dirt and unwanted residues from the skin but they will also remove some of the natural oils that skin produces for its own protection. The shower gel therefore needs to contain moisturisers and oils (emollients) to replace those that are lost, and since these do not naturally blend with surfactants, which prefer to dissolve in water, the shower gel will need to contain emulsifiers.

It used to be thought that cleansing products should be alkaline because this removed the grease to which dirt clings, and soaps were of this type. However, soaps also removed more of the natural oils leaving the skin feeling dry and itchy.

© Shutterstock

Today the trend is to have cleaning agents that are slightly acidic to prevent this happening.

Products for the bathroom need another essential ingredient, namely antimicrobials, to ensure that germs don't breed in the product as these might cause skin infections. You might wonder as you read the formulations: why are there so many different kinds of these in a single product? The answer is that many of the ingredients have themselves to be protected by suitable antimicrobial agents during storage and transport, and these then become part of the final product.

The other ingredients in shower gels are fragrances to make the product smell nice and dyes to give it a pleasing colour. The dyes are generally indicated by colour index (CI) number.

As you are no doubt aware, there are lots of shower gels available – one website lists more than 500 – but all have to contain the same basic chemicals if they are to deliver the result we expect: a clean body and fresh feel. Five products are included in this section; three designed for women and two for men.

## Dove® Deeply Nourishing Body Wash

*The Chemistry*

All shower gels want to offer a little something different and this one says it contains 'Nutrium Moisture™'. This is said to be deeply absorbing and nourishing, and it consists of sodium lauroyl isethionate, sodium isethionate, sodium stearate, soybean oil, and glycerine, which together not only clean but moisturise the skin and replace lost natural oil.

*Contents*

| | |
|---|---|
| Aqua | Sodium chloride |
| Glycerin | Stearic acid |
| Cocamidopropyl betaine | Sodium palmitate |
| Sodium laureth sulfate | Sodium lauroyl isethionate |
| Sodium hydroxypropyl starch phosphate | Sodium isethionate |
| *Helianthus annuus* hybrid oil | Sodium stearate |
| Sodium cocoyl glycinate | Tetrasodium EDTA |
| Lauric acid | Sodium palm kernelate |
| Hydrogenated soybean oil | Tetrasodium etidronate |
| Parfum | Citric acid |
| Guar hydroxypropyltrimonium chloride | DMDM hydantoin |
| | Sodium benzoate |
| | Methylisothiazolinone |

BHT                                          Citronellol
Zinc oxide                                   Hexyl cinnamal
Alumina                                      Limonene
Benzyl alcohol                               Linalool
Butylphenyl methylpropional                  CI 77891

**Aqua** is water and acts as a solvent.

*Surfactants*

These have a dual purpose: to cleanse the skin and to stabilise the shower gel. Most of them come from renewable natural sources.

**Cocamidopropyl betaine** is a non-ionic surfactant that foams very easily.

**Sodium laureth sulfate** is an anionic surfactant.

**Sodium cocoyl glycinate** is both a surfactant and a skin conditioner as is **sodium lauroyl isethionate**.

**Sodium palmitate** acts as a gentle surfactant, emulsifier, and viscosity regulator.

**Sodium isethionate** is an anionic surfactant that foams well.

**Sodium stearate** is an anionic surfactant and emulsifier, and it makes the gel viscous.

**Sodium palm kernelate** is an anionic surfactant.

*Emollients and skin conditioners*

**Glycerin** acts to moisturise the skin.

*Helianthus annuus* **hybrid oil** is sunflower oil.

**Hydrogenated soybean oil** is the plant oil that has had the unsaturated bonds of its fatty acid chains converted to saturated ones, thereby making it more stable.

*Emulsifiers and Thickeners*

**Lauric acid** and **stearic acid** are long chain saturated fatty acids that act as emulsifiers.

**Sodium chloride** is common salt, which acts to thicken the gel.

*Colourants*

**Alumina** makes the gel opaque.

**CI 77891** is the white pigment titanium dioxide, which makes the product opaque.

**Sodium hydroxypropyl starch phosphate** acts as a mild abrasive to remove dead skin cells.

**Parfum** is a mixture of fragrances, some natural, some artificial, and it may contain items that some individuals are sensitive to. These are identified as **butylphenyl methylpropional**, which is an artificial fragrance with a strong floral note; **citronellol**, which smells of geraniums and roses; **hexyl cinnamal**, which smells of camomile; **limonene**, which smells of oranges; and **linalool**, which has a floral spicy note.

**Guar hydroxypropyltrimonium chloride** is an antistatic agent, mostly used in shampoos.

**Tetrasodium EDTA** and **tetrasodium etidronate** deactivate the calcium ions of hard water so they don't interfere with the surfactants.

**Citric acid** ensures the gel is slightly acidic.

**DMDM hydantoin, sodium benzoate, methylisothiazolinone, BHT** (butylated hydroxytoluene), and **benzyl alcohol** are powerful antimicrobial agents that have been added to the various ingredients.

**Zinc oxide** is thought to be beneficial to the skin.

## Imperial Leather® Limited Edition Shower Crème

This is basically a mixture of surfactants and natural oils, and the product is slightly alkaline.

### Contents

Aqua (water)
Sodium laureth sulfate
Cocamidopropyl betaine
Acrylate/steareth-20 methacrylate
  crosspolymer
Glycerin
PEG-6
Caprilic/capric glycerides
Parfum
*Salix alba* (willow) bark extract
*Helianthus annuus* (sunflower)
  seed oil
*Simondsia chinensis* (jojoba)
  seed oil
Lithium magnesium sodium
  silicate
Tetrasodium pyrophosphate
Methylisothiazolinone

Magnesium chloride
Magnesium nitrate
Tetrasodium glutamate diacetate
Sodium benzoate
Sodium hydroxide
Benzotriazolyl dodecyl *p*-cresol
Butyl methoxybenzoylmethane
Gelatin
*Acacia senegal* gum
Silica
Mica
Xanthan gum
Hexyl cinnamal
Linalool
CI77891 (titanium dioxide)
CI77007
CI75470

**Aqua** is water and acts as a solvent.

## Surfactants

**Sodium laureth sulfate** is an anionic surfactant, while **cocamidopropyl betaine** is a non-ionic surfactant; both are gentle on the skin and foam easily.

## Emollients and skin conditioners

**Glycerin** and **caprilic/capric glycerides** soften the skin, while **gelatin** is the protein of collagen and should also benefit the skin.

**PEG-6** is a humectant and will moisturise the skin.

The natural oils are *Salix alba* **extract,** which is extracted from willow bark; *Helianthus annuus,* which comes from sunflowers seeds; and *Simondsia chinensis*, which is extracted from jojoba seeds.

## Emulsifiers and Thickeners

**Acrylate/steareth-20 methacrylate crosspolymer,** *Acacia senegal* **gum** (also known as gum Arabic), and **xanthan gum** act as thickeners and emulsifiers.

## Colourants

**Lithium magnesium sodium silicate** is talc-like clay, which gives the gel an attractive opalescence. **CI77891** is the white pigment titanium dioxide, **CI77007** is ultramarine, and **CI 75470** is carmine.

**Parfum** is a mixture of fragrances, some natural, some artificial, and it may contain items that some individuals are sensitive to. These are identified as **hexyl cinnamal,** which smells of camomile, and **linalool,** which has a pleasing floral note with a hint of spiciness.

**Tetrasodium pyrophosphate** and **tetrasodium glutamate diacetate** are chelating agents that combine with any calcium in the water so it doesn't interfere with the surfactants.

**Methylisothiazolinone** and its related molecule **methylchloroisothiazolinone** are powerful germ killers, as is **sodium benzoate,** which is a naturally occurring chemical.

**Magnesium chloride** and **magnesium nitrate** provide magnesium ions, which some believe have a beneficial effect on the skin.

**Silica** (silicon dioxide, sand) is there as an abrasive to help remove dead skin cells, as is **mica** (magnesium aluminium silicate), which is a gentle abrasive.

**Sodium hydroxide,** formula NaOH, is added to keep the pH above 7. Slightly alkaline wash water is known to be more effective at cleaning than slightly acidic conditions.

**Benzotriazolyl dodecyl *p*-cresol** and **butyl methoxybenzoylmethane** are antioxidants. These are known as photostabilizers and are used in some sunscreens.

## Rituals® Zensation Foaming Shower Gel

*The Chemistry*

This product comes out of its pressurised container as a gel but after it has been spread on the skin it quickly, almost magically, transforms itself into foaming lather. This effect is achieved by one of its ingredients, isopentane, which is a hydrocarbon with a boiling point of 28 °C. The warmth of the skin causes it to evaporate, creating foam as it does.

*Contents*

Aqua/water
Sodium laureth sulfate
Isopentane
Cocamidopropyl betaine
Sorbitol
Isopropyl palmitate
Parfum/fragrance,
PEG-120 methyl glucose
  dioleate
Isobutane
Guar hydroxypropyltrimonium
  chloride
*Prunus cerasus* (bitter cherry) extract
Butylene glycol
Alpha isomethyl ionone
Benzyl alcohol
Benzyl salicylate
Butylphenyl methylpropional
Cinnamyl alcohol
Citronellol
Coumarin
Geraniol
Hexyl cinnamal
Linalool

**Aqua** is water and acts as a solvent.

*Surfactants*

**Sodium laureth sulfate** is an anionic surfactant that is made from coconut oil and it is gentle and lathers well, while **cocamidopropyl betaine** is a non-ionic surfactant that also foams very easily.

### Emollients and skin conditioners

**Guar hydroxypropyltrimonium chloride** is a derivative of guar gum that is water soluble and it deposits itself on the skin as a conditioner.

*Prunus cerasus* **(bitter cherry) extract** is a natural oil.

### Emulsifiers and Thickeners

**Sorbitol** and **isopropyl palmitate** act as a thickening agents.

**PEG-120 methyl glucose dioleate** acts as a thickener and emulsifier.

**Isopentane** evaporates, thereby causing the product to produce a layer of foam.

**Isobutane** is a hydrocarbon gas with a boiling point of –12 °C. This is the propellant gas, which is under pressure in the canister and forces out the content through a nozzle when the pressure is released.

**Parfum.** Perfume raw material is a complex mixture, some items of which have to be identified on a label because there are people who may be sensitive to them. The ones so identified here are **butylphenyl methylpropional**, which has a fresh outdoor smell; **cinnamyl alcohol**, which smells of hyacinth; **citronellol**, which is the main fragrance of geraniums and roses; **coumarin**, which has the scent of new-mown hay; **geraniol**, which smells of roses; **hexyl cinnamal**, which smells of camomile; and **linalool**, which has a pleasing floral note with a hint of spiciness.

**Benzyl alcohol** and **benzyl salicylate** are preservatives.

### Radox® for Men

This is a typical shower gel, although available with different fragrances.

### Contents

Aqua
Sodium laureth sulfate
Cocamidopropyl betaine
Sodium chloride
Glycerin
Allantoin
Maris sal

*Foeniculum vulgare* extract
Parfum
Citric acid
Sodium lactate
Polyquaternium-7
Linalool
Limonene

Benzyl salicylate
Citronellol
Butylphenyl methylproprional
Hydroxyisohexyl 3-cyclohexane
  carboxyaldehyde
Trideceth-9

PEG-40 hydrogenated castor oil
Sodium benzoate
CI 42053
CI 17200

**Aqua** is water and acts as a solvent.

*Surfactants*

**Sodium laureth sulfate** is an anionic surfactant that is gentle
and lathers well, while **cocamidopropyl betaine** is a non-
ionic surfactant that also foams easily.

*Emollients and skin conditioners*

**Glycerin** keeps the skin moist.

*Foeniculum vulgare* **extract** is from the common herb fennel
and has a slight aniseed smell. It is thought to have a
beneficial effect on the skin.

**Sodium lactate** acts as a moisturiser and makes the outer layer
of skin softer.

**Polyquaternium-7** is a long-chain molecule with a positive end
that is attracted to the negatively charged surface of skin
leaving a film that makes it feel softer.

*Emulsifiers and Thickeners*

**Sodium chloride** is common salt, as is **maris sal**, which is
simply sea salt. These act as thickeners.

**Trideceth-9** and **PEG-40 hydrogenated castor oil** are
emulsifiers.

*Colourants*

**CI 42053** is green and **CI 17200** is red.

**Allantoin** is said to assist the removal of dead cells from the
surface of the skin thereby leaving it feeling smooth to the
touch. It also possibly rehydrates living skin cells.

**Parfum.** Perfume raw material is a complex mixture, some
items of which have to be identified on a label because there
are people who may be sensitive to them. The ones so
identified here are **linalool**, which has a floral smell with a
hint of spice; **limonene**, which comes from the peel of

oranges; and **citronellol**, which is the main fragrance of geraniums and roses. The synthetic fragrances identified are **butylphenyl methylproprional** and **hydroxylisohexyl 3-cyclohexane carboxyaldehyde**, which smell of lily-of-the-valley.

**Citric acid** keeps the pH of the gel below 7, giving it a fresher feel on the skin.

**Benzyl salicylate** is a fixative; in other words it is there to help the fragrance molecules to blend in with the other ingredients. Of itself, it has almost no odour.

**Sodium benzoate** is a powerful germicide and is particularly effective under acid conditions.

## Radox® for Men, Mint and Tea Tree Shower Gel

This includes chemicals that make the skin tingle. The active ingredients are like those of the normal Radox® for Men, but with fewer fragrances. It now has ingredients that tone up the skin.

*Contents*

| | |
|---|---|
| Aqua | Polyquaternium-7 |
| Sodium laureth sulfate | Tetrasodium EDTA |
| Cocamidopropyl betaine | Limonene |
| Parfum | Coumarin |
| Sodium chloride | Hexyl cinnamal |
| *Mentha arvensis* leaf oil | Styrene/acrylates copolymer |
| *Melaleuca alternifolia* leaf oil | Sodium benzoate |
| Citric acid | CI 42090 |
| Sodium lactate | CI 47005 |

**Aqua** is water and acts as a solvent.

*Surfactants*

**Sodium laureth sulfate** is an anionic surfactant that is gentle and lathers well, while **cocamidopropyl betaine** is a nonionic surfactant that also foams easily.

*Emollients*

**Sodium lactate** acts as a moisturiser and makes the outer layer of skin softer.

**Polyquaternium-7** is a long-chain molecule with a positive end that is attracted to the negatively charged surface of skin leaving a film that makes it feel softer.

*Emulsifiers and Thickeners*

**Sodium chloride** is common salt. It acts as a thickener.

*Colourants*

**CI 42090** is blue and **CI 47005** is yellow, and together they produce a green colour.

*Mentha arvensis* **leaf oil** comes from the mint plant, and was once thought to relieve aches and pains when rubbed on the skin.

*Melaleuca alternifolia* **leaf oil** is tea tree oil and has around 100 molecular components. These include terpinen-4-ol, and it is this that provides a tingling sensation on the skin as well as having antimicrobial properties.

**Tetrasodium EDTA** is a chelating agent to prevent any calcium in the water from interfering with the surfactants.

**Citric acid** keeps the pH of the gel below 7, giving it a fresher feel on the skin.

**Sodium benzoate** is a powerful germicide and is particularly effective under acid conditions.

**Styrene/acrylates copolymer** is used to stabilise the gel.

**Parfum.** Perfume raw material is a complex mixture, some items of which have to be identified on a label because there are people who may be sensitive to them. The ones so identified here are **limonene**, which comes from the peel of oranges; **coumarin**, which has the scent of new-mown hay; and **hexyl cinnamal**, which smells of camomile.

Other shower gels are available for both men and women under the brand names of Sanex®, Nivea®, Palmolive®, Lynx® *etc.*

## 2. BATH PRODUCTS

*The Chemistry*

A bath is essential for washing babies or young children. (It can be a relaxing experience for adults as well, although it does use much more water than taking a shower.) One thing a parent looks for when choosing a product for bathing a young child is something that does not affect the eyes and this kind of bath product generally avoids all but the most essential ingredients.

For adults there are almost as many products to add to the bath as there are shower gels, most of them varying the fragrances they contain although some contain herbs that are supposed to offer benefits and some even claim to 'detox' the body.

Here we will look at one product for bathing a baby, and one for adults.

### Johnson's® Baby Bath

This product claims it contains nothing that would sting if it got into a baby's eyes.

*Contents*

| | |
|---|---|
| Aqua | PEG-80 sorbitan laurate |
| Coco-glucoside | Polysorbate-20 |
| Sodium laureth sulfate | Sodium glycolate |
| Sodium lauroamphoacetate | Phenoxyethanol |
| Sodium chloride | Sodium benzoate |
| Citric acid | Parfum |
| Cocamidopropyl betaine | CI 42090 |
| PEG-150 distearate | |

**Aqua** is water and acts as a solvent.

**Coco-glucoside, sodium laureth sulfate, cocamidopropyl betaine** and **polysorbate-20** are mild surfactants that lather well. **Polysorbate-20** also ensures the various ingredients remain smoothly blended.

**Sodium chloride** is common salt and this thickens the formulation.

**Citric acid** keeps the pH below 7 and it also acts as a chelating agent.

**PEG-150 distearate** and **PEG-80 sorbitan laurate** are humectants.

**Sodium glycolate** is a buffering agent to keep the pH stable.

**Sodium lauroamphoacetate** is a foaming agent.

**Phenoxyethanol** and **sodium benzoate** prevent microbial contamination of the product.

**Parfum** is unspecified, but it has a pleasant floral note and it does not contain any of the fragrances that would otherwise have to be identified.

**CI 42090** is blue.


**Radian® B Mineral Bath**

Although this product is marketed as a mineral bath, it is not clear from the packaging what this mineral is, apart from added salt. However, because the product lacks a sequestering agent, this means that the minerals normally present in hard water, namely calcium and magnesium, will not be deactivated.

*Contents*

| | |
|---|---|
| Aqua | *Anthemis noblis* |
| Sodium laureth sulfate | *Malva sylvestris* |
| Cocamidopropyl betaine | *Melissa officinalis* |
| Sodium chloride | Geraniol |
| Polysorbate-20 | Limonene |
| Perfume | Linalool |
| Benzophenone-3 | CI 42090 |
| Benzyl alcohol | CI 19140 |
| Methylchloroisothiazolinone | |

**Aqua** is water and acts as a solvent.

**Sodium laureth sulfate** is an anionic surfactant that is made from coconut oil. It is gentle and lathers well. **Cocamidopropyl betaine** is a non-ionic surfactant that also foams easily.

**Sodium chloride** is common salt and here it provides the mineral component as well as acting as a thickener.

**Polysorbate-20** ensures the various ingredients blend smoothly together.

**Benzophenone-3** protects the product against UV rays.

**Benzyl alcohol** acts partly as a solvent for other ingredients but also prevents the growth of bacteria.

**Methylchloroisothiazolinone** is a more potent antimicrobial agent.

***Anthemis noblis*** is camomile, and ***Melissa officinalis*** is lemon balm. Camomile is an emollient and a skin conditioner, while lemon balm is a natural oil of long-standing use in so-called toilet water.

***Malva sylvestris*** is common mallow and is thought to tone up the skin.

**Parfum** Perfume raw material is a complex mixture, some items of which have to be identified on a label because there are people who may be sensitive to them. The ones so identified here are **geraniol**, which smells of roses; **limonene**, which comes from the peel of oranges; and **linalool**, which smells of lavender with a hint of spiciness.

**CI 42090** is blue and **CI 19140** is yellow.

Other bath products are available under the brand names of Dove®, Radox®, Cussons®, *etc.*

### 3.   SHAMPOOS AND CONDITIONERS

*The Chemistry*

There are more than 400 kinds of shampoo on sale in the UK. All contain surfactants and these have to be gentle so as not to remove too much natural oil from the hair or it will lose its glossy look. Some shampoos include silicones to protect the scalp and hair, although most now deliberately avoid them, probably because of bad publicity surrounding the use of silicones in breast implants.

Here we look at two shampoos, a regular one, and one that is designed to remove dandruff and prevent it recurring.


**Pantene Pro-V®**

The pack says that this is for fine hair, producing volume and body, and also says it is a silicone-free shampoo.

© Shutterstock

## Contents

Aqua
Sodium lauryl sulfate
Sodium laureth sulfate
Cocamidopropyl betaine
Sodium citrate
Sodium xylenesulfonate
Sodium chloride
Parfum
Citric acid
Sodium benzoate
Hydroxypropyl methylcellulose
Tetrasodium EDTA
Butylphenyl methylpropional
Panthenol
Panthenyl ethyl ether
Linalool
Hexyl cinnamal
Limonene
Benzyl salicylate
Magnesium nitrate
Methylchloroisothiazolinone
Magnesium chloride
Methylisothiazolinone

**Aqua** is water and acts as a solvent.

**Sodium lauryl sulfate** and **sodium laureth sulfate** are anionic surfactants while **cocamidopropyl betaine** is a non-ionic surfactant. All are gentle, and foam easily.

**Sodium citrate**, along with **citric acid**, acts as a buffer to keep the shampoo slightly acidic and the pH stable.

**Sodium xylenesulfonate** and **hydroxypropyl methylcellulose** are emulsifiers.

**Sodium chloride** increases the viscosity.

**Parfum.** Perfume raw material is a complex mixture, some items of which have to be identified on a label because there are people who may be sensitive to them. The ones so identified here are **butylphenyl methylpropional**, which has a fresh outdoor smell; **linalool**, which has a pleasing floral note with a hint of spiciness; **hexyl cinnamal**, which smells of camomile; and **limonene**, which comes from the peel of oranges.

**Sodium benzoate** is a powerful germicide and is particularly effective under acid conditions.

**Tetrasodium EDTA** is a chelating agent and removes calcium hardness from the water.

**Panthenol** is similar to vitamin B5 and is attracted to the hair and makes it shine.

**Panthenyl ethyl ether** is an antistatic agent that makes hair manageable.

**Magnesium nitrate** and **magnesium chloride** provide magnesium ions, which some believe have a beneficial effect on the skin.

**Methylchloroisothiazolinone** and **methylisothiazolinone** are powerful antimicrobials.

**Benzyl salicylate** is a fixative and helps the fragrance molecules to blend in with the other ingredients. Of itself, it has almost no odour.

## Head and Shoulders®

In addition to washing the hair, this will kill the fungus that causes dandruff.

*Contents*

Aqua
Sodium laureth sulfate
Sodium lauryl sulfate
Cocamide MEA
Zinc carbonate
Glycerol distearate
Sodium chloride
Zinc pyrithione
Dimethicone
Cetyl alcohol
Guar hydroxypropyltrimonium
  chloride
Sodium xylenesulfonate
Magnesium sulfate
Sodium benzoate
Ammonium laureth sulfate
Butylphenyl methylpropional
Sodium diethylenetriamine
  pentamethylene phosphonate

Magnesium carbonate hydroxide
Hexyl cinnamal
Benzyl alcohol
Etidronic acid
Hydroxyisohexyl 3-cyclohexene
  carboxaldehyde
Limonene
Citronellol
Paraffinum liquidum
Sodium polynaphthalenesulfonate
Methylchloroisothiazolinone
DMDM hydantoin
Disodium EDTA
Tetrasodium EDTA
Methylisothiazolinone
CI 42090
CI 6073

**Aqua** is water and acts as a solvent.

**Sodium laureth sulfate, sodium lauryl sulfate**, and **ammonium laureth sulfate** are anionic surfactants that lift off dirt and grease. They are gentle and produce a rich lather.

**Cocamide MEA** is a non-ionic surfactant and also has excellent foaming properties.

**Zinc carbonate** makes the shampoo opaque and may have some skin protection factor.

**Glycerol distearate** is an antistatic agent to prevent the hair from having a fly-away look.

**Sodium chloride** thickens the shampoo.

The active agents are **zinc pyrithione** and **DMDM hydantoin**, which are powerful antimicrobial agents that kill the fungus that causes dandruff.

**Dimethicone** is a colourless silicone oil that leaves a protective, water-repelling layer on the surface of the skin and hair and helps them retain moisture.

**Cetyl alcohol** acts as an emollient.

**Guar hydroxypropyl methylcellulose** and **sodium xylenesulfonate** act as emulsifiers.

**Magnesium sulfate** and **magnesium carbonate hydroxide** provide magnesium, which is thought to be good for skin and hair.

**Sodium benzoate** is a preservative and is particularly effective under acid conditions.

The perfumes here are a complex mixture, some items of which have to be identified on a label because there are people who may be sensitive to them. The ones so identified here are **limonene**, which is extracted from orange peel; **butylphenyl methylpropional**, which has a fresh outdoor smell; **hexyl cinnamal**, which smells of camomile; **citronellol**, which is the fragrance of geraniums and roses; and **hydroxyisohexyl 3-cyclohexene carboxaldehyde**, which is an artificial fragrance that smells like lily of the valley.

**Sodium diethylenetriamine pentamethylene phosphonate**, **etidronic acid**, **disodium EDTA**, and **tetrasodium EDTA** act as chelants to neutralise the action of the calcium in hard water.

**Benzyl alcohol** acts partly as a solvent and a preservative.

**Paraffinum liquidum** (liquid paraffin) is a hydrocarbon oil that gives the hair a glossy look.

**Sodium polynaphthalenesulfonate** is an emulsifier and it has some surfactant activity.

**Methylchloroisothiazolinone** and **methylisothiazolinone** are antimicrobials.

**CI 42090** is blue and **CI 60730** is violet.

## L'Oréal® Elvive Nutri-Gloss® Conditioner

*The Chemistry*

Conditioners are used to make the hair smoother, softer, and stronger and to ensure it does not tangle or become brittle and break. Conditioners are attracted to hair, which is negatively charged, by being positively charged and, in so doing, remove static electricity from the hair.

*Contents*

**Aqua** is water and acts as a solvent.

**Behentrimonium chloride** is attracted to the strands of hair by virtue of its positive charge and then it acts as an antistatic agent. It also has biocidal properties. A similar ingredient is **stearamidopropyl dimethylamine**, which is also attracted to the hair by virtue of its two nitrogen atoms.

L'Oreal® Elvive contains four alcohols that act as lubricants, emollients, and emulsifiers. They also help give the hair more gloss and stop it becoming dry and brittle. They are **stearyl alcohol, cetyl alcohol, octyldodecanol**, and **myristyl alcohol**.

**Octyldodecanol** makes the product opaque.

**Palm oil** comes from the African palm tree *Elaeis guineensis*, and **jojoba seed oil** comes from the *Simmondsia chinensis* shrub. They coat the hair and give it an attractive gloss.

**Glycerin** acts to ensure all the ingredients blend nicely and that water is not lost.

**Fragrance parfum** Perfume raw material is a complex mixture, some items of which have to be identified on a label because there are people who may be sensitive to them. The ones so identified here are **limonene**, which comes from orange peel; **hexyl cinnamal**, which smells of camomile; while **linalool** has a pleasing floral note with a hint of spiciness; and **citronellol** smells of geraniums and roses. Other, artificial, fragrances are **butylphenyl methylpropional** which has a fresh outdoor smell and **alpha-isomethyl ionone** whose smell is described as 'clean.'

**Methylparaben** and **chlorhexidine dihydrochloride** are antimicrobial agents.

**Hydroxyethylcellulose** acts as a humectant and increases the viscosity and lubricity.

**Citric acid** is used to adjust the pH so that the conditioner is slightly acidic and so will not remove natural oils from the skin.

**Hydrolyzed conchiolin protein** comes from snail shells and it has a pearly lustre. This gives the conditioner its light reflecting qualities, and is unique to this product.

**CI 17200** is red.

Other conditioners are available under the brand names of Dove®, Pantene®, Revlon®, TRESemmé®, Nivea®, Neal's Yard®, John Frieda®, Molton Brown®, *etc.*

## 4.   TOOTHPASTE

*The Chemistry*

Tooth enamel is a form of calcium phosphate known as hydroxyapatite and this protects the dentine, which has nerve fibres. If dentine is exposed because the enamel has been damaged through decay, or the gums have receded, then the teeth become painfully sensitive to cold and hot liquids.

Oral bacteria are present in dental plaque, which is a biofilm that forms on the teeth. The bacteria can convert sugar to acid and thereby corrode the enamel. Plaque can be removed by brushing the teeth with a toothpaste that consists of a surfactant and an abrasive. The toothpaste can be made pleasant to taste by the inclusion of an artificial (non-sugar) sweetener.

Converting the hydroxyapatite to fluoroapatite makes it stronger and more acid resistant and this can be done by including fluoride in the toothpaste. In formulations designed for children this needs to be less than in adult formulations because children have a tendency to swallow toothpaste. Adult fluoride toothpaste formulations generally have around 1500 ppm fluoride whereas for children it is half this amount.

Some toothpastes are formulated to plug tiny holes in the enamel.

© Shutterstock

## Colgate® Fluoride Toothpaste

*Contents*

Dicalcium phosphate dihydrate
Aqua
Glycerin
Sodium lauryl sulfate
Cellulose gum
Aroma
Sodium monofluorophosphate
Tetrasodium pyrophosphate

Sodium saccharin
Sodium fluoride
Calcium glycerophosphate
Limonene
Contains: Monofluorophosphate
(1000 ppm F⁻)
Sodium Fluoride (450 ppm F⁻)

**Dicalcium phosphate dihydrate** is a mild abrasive.

**Aqua** is water and acts as a solvent.

**Glycerin** acts as a humectant.

**Sodium lauryl sulfate** is an anionic surfactant.

**Cellulose gum** acts to hold all the ingredients together in a smooth paste.

**Aroma** is unspecified.

**Sodium monofluorophosphate** provides fluoride in the form in which it is present in tooth enamel, and the fluoride component is 1000 ppm.

**Tetrasodium pyrophosphate** is an emulsifier and thickening agent.

**Sodium saccharin** is an artificial sweetener.

**Sodium fluoride** provides some of the fluoride (450 ppm).

**Calcium glycerophosphate** is thought to be more readily absorbed into tooth surfaces than simple phosphate.

**Limonene** is a hydrocarbon oil extracted from orange peel.

## Colgate Sensitive Pro-Relief®

*Contents*

Arginine bicarbonate (8%)
Calcium carbonate
Aqua (water)
Sorbitol
Sodium lauryl sulfate
Sodium monofluorophosphate
(1450 ppm F⁻)
Aroma

Sodium silicate
Cellulose gum
Sodium bicarbonate
Titanium dioxide
Potassium acesulfame
Xanthan gum
Sucralose

**Aqua** is water and acts as a solvent.

**Arginine bicarbonate** is an amino acid compound and is a component of proteins. In conjunction with **calcium carbonate** it can plug holes in tooth enamel and dentine.

**Sorbitol** is a humectant. It is a sweet tasting sugar alcohol but, unlike sugar, it does not cause tooth decay due to microbial action that forms acids.

**Sodium lauryl sulfate** is a foaming surfactant.

**Sodium monofluorophosphate** provides fluoride in the form in which it is present in tooth enamel, and the fluoride component is 1450 ppm.

**Sodium silicate** is a gentle abrasive.

**Cellulose gum** and **xanthan gum** hold all the other ingredients together as a homogeneous paste.

**Titanium dioxide** is an intensely white pigment.

**Potassium acesulfame** and **sucralose** are artificial sweeteners.

### Sensodyne® Original Toothpaste

*The Chemistry*

This contains a strontium salt that can replace calcium in the tooth enamel and help plug any minute holes, so preventing access to the underlying nerves of the dentine.

*Contents*

| | |
|---|---|
| Strontium chloride hexahydrate (10%) | Polyoxyl 40 stearate |
| Water | Titanium dioxide |
| Glycerol | Sodium saccharin |
| Sorbitol | Erythosine (E127) |
| Calcium carbonate | Spearmint oil |
| Hydroxyethyl cellulose | Peppermint oil |
| Colloidal anhydrous silica | Levomenthol |
| Sodium methyl cocoyl taurate | Methyl salicylate |
| | Oil of cassia |

**Strontium chloride hexahydrate** reduces tooth sensitivity by repairing damage to the tooth enamel.

**Glycerol** and **sorbitol** act as humectants.

**Calcium carbonate** and **colloidal anhydrous silica** act as gentle abrasives.

**Hydroxyethyl cellulose** is a polymer designed to give the toothpaste the right consistency.

**Sodium methyl cocoyl taurate** is a mild surfactant.

**Polyoxyl 40 stearate** is a non-ionic surfactant.

**Titanium dioxide** is a white pigment.

**Sodium saccharin** is an artificial sweetener.

**Erythosine** gives the toothpaste its pink coloration.

**Spearmint oil** and **peppermint oil** provide a fresh feel to the mouth.

**Levomenthol** and **menthol** also occurs naturally in peppermint oil. They trigger the cold-sensitive receptors in the mouth, producing a fresh feel.

**Methyl salicylate** is also known as oil of wintergreen and has a strong characteristic aroma.

**Oil of cassia** comes from the bark of a Southeast Asian tree *Cinnamomum cassia*.

There are more than 100 varieties of toothpaste on sale in the UK and other popular brands are Oral-B®, Aquafresh®, and Macleans®.

## 5.  DENTURE CLEANER TABLETS

### Steradent® Active Plus

*The Chemistry*

Just as with normal teeth, dentures need cleaning regularly to remove the plaque that forms on their surface. It is possible to brush this off with normal toothpaste but an easier way to clean them is to leave them soaking overnight in a solution designed to remove the unwanted film of bacteria and food particles that forms during the day. This is easily done by placing them in water along with a tablet, such as Steradent®. As the ingredients in the tablet dissolve, the sodium bicarbonate and other carbonates react chemically with the citric acid. This releases bubbles of carbon dioxide, which keeps the liquid stirred and so maximises contact of the other chemicals with the dentures to ensure they are cleaned and sterilised by the other ingredients that are present.

*Contents*

| | |
|---|---|
| Sodium bicarbonate | Sodium cocoyl isethionate |
| Sodium sulfate | Aroma |
| Citric acid | Glucose |
| Sodium carbonate peroxide | Sodium chloride |
| Sodium carbonate | Methenamine |
| Potassium carbonate | Sulfamic acid |
| Malic acid | Cetylpyridinium chloride |
| PEG-150 | CI 73015 |
| PEG-90 | |

**Sodium bicarbonate** reacts with the **citric acid** to generate bubbles of carbon dioxide. **Sodium carbonate** and **potassium carbonate** react in the same manner but at a slower pace generating bubbles over a longer period.

**Sodium sulfate, glucose** and **sodium chloride** appear to be mainly there to add bulk to the tablets.

**Sodium carbonate peroxide** is a combination of sodium carbonate and hydrogen peroxide. It releases hydrogen peroxide when dissolved in water and this is a powerful bleaching agent.

**Malic acid** is also used by dentists to clean teeth before they are filled.

**PEG-150** and **PEG-90** are polyethylene glycol polymers with long chains. These make the water slightly more viscous, which helps keep the bubbles in the solution for longer. They also bind together the other components in the tablet.

**Sodium cocoyl isethionate** is a mild anionic surfactant that is a solid and readily dissolves in water.

**Aroma.** There is no ingredient needing to be identified that provides the aroma, but these tablets have an antiseptic and mint-like smell.

**Methenamine** is an antibiotic agent.

**Sulfamic acid** is a strong acid but unlike many common acids it is a white crystalline solid. **Cetylpyridinium chloride** is an antibacterial agent

**CI 73015** provides the pink colour.

## 6.   MOUTHWASH

*The Chemistry*

The microbes that live in the mouth are generally harmless, but they can be responsible for the release of unpleasant smelling chemicals, which are mostly sulfur-containing molecules such as those that cause bad breath (halitosis). Cleaning the plaque from the teeth will generally reduce the colonies of microorganisms to levels that ensure your breath is not offensive. To be certain that your breath is fresh, there are mouthwashes with antimicrobial chemicals that will wipe out most of the residual bacteria, including those lurking where the toothbrush cannot reach.

## Listerine® Original

Listerine® is one of the best known brands of mouthwash. It was formulated in 1879 by Joseph Lawrence and Jordan Lambert of St Louis, Missouri, to be used as a surgical antiseptic. They named it after Sir Joseph Lister (1827–1912) who first introduced antiseptic surgery in 1867 at Glasgow Royal Infirmary. Today there are several varieties of Listerine, including alcohol-free ones, ones that whiten teeth, and ones with various flavours. The original formulation is as follows:

*Contents*

| | |
|---|---|
| Aqua | Methyl salicylate |
| Alcohol | Thymol |
| Benzoic acid | Menthol |
| Poloxamer 407 | Sodium benzoate |
| Eucalyptol | Caramel |

**Aqua** and **alcohol** (which comprises 27% of the solution) are solvents for the various ingredients.

**Benzoic acid** and its sodium salt, **sodium benzoate**, are there to kill bacteria, including any that might find their way into the Listerine bottle itself.

**Poloxamer 407** is a non-ionic surfactant.

**Eucalyptol** (of which there is 0.092%) comes from eucalyptus oil and has a refreshing aroma.

**Methyl salicylate** (of which there is 0.060%) is also known as oil of wintergreen and is what gives traditional embrocation its characteristic odour.

**Thymol** (of which there is 0.064%) has strong antiseptic properties.

**Menthol** (of which there is 0.042%) occurs naturally in peppermint oil and it triggers the cold-sensitive receptors in the mouth, producing a fresh feel.

**Caramel** is polymerised sugar, and this gives the solution flavour and a pale yellow colour.

### Corsodyl® Medicated Mint Mouthwash

*The Chemistry*

This mouthwash also heals inflamed gums and mouth ulcers. If using a toothbrush is difficult or not recommended, such as after surgery involving the mouth, then this product will prevent plaque forming on the teeth.

*Contents*

| | |
|---|---|
| Water | Macroglycerol hydroxystearate |
| Chlorhexidine digluconate (0.2%) | Sorbitol |
| Ethanol (7%) | Peppermint oil |

**Water** and **ethanol** (which as alcohol accounts for 7% of the product) are solvents.

**Chlorhexidine digluconate (0.2%)** is an antiseptic that is active against bacteria and fungi, and it persists in the mouth for longer than other antiseptics.

**Macroglycerol hydroxystearate** is an emulsifier that helps the antiseptic agent to dissolve.

**Sorbitol** is a natural sweetener that cannot cause tooth decay.

**Peppermint oil** provides the minty flavour.

Other mouthwashes are Dentyl Active®, Aquafresh®, Colgate® and Oral-B® and these contain similar ingredients to those mentioned above for Listerine®.

## 7.   HAIR DYES

*The Chemistry*

There are three kinds of hair dye. There are those that are
temporary and are easily washed out, and these are the kind of
colourant you might apply for a fancy dress party or a stage
performance. Then there are those that dye the hair as
you would dye a fabric, and these are longer lasting but will
eventually fade with each subsequent wash with shampoo.
Finally there are those that are permanent, although they will
of course eventually grow out. These permanent hair dyes
are inside the hair itself, and they require three application
steps:

1. The first opens up the outer layer of the hair, the cuticles, so
   that other chemicals can gain access.
2. The second applies the dye precursors, which penetrate the
   hair and once inside it they react chemically to produce the
   desired colour. This requires hydrogen peroxide, the so-
   called colour activator, which enables the chemicals to form
   the dye molecules. The skill of the cosmetic chemists has
   been to devise a combination that produces the shade the
   customer desires.
3. The third step closes the cuticles so the dye molecules
   cannot escape. It also repairs any damage to the hair that
   the dyeing procedure might have caused, and leaves it
   feeling nice and looking shiny.

### Clairol Nice'n'Easy™ Hair Dye Permanent Natural Medium Mahogany Brown

There are many shade of hair dye. The one here provides a
golden auburn colour.

### Step 1: Colour Blend Formula

This cleans the hair and then the ammonium hydroxide opens
its cuticles to allow the dye-forming molecules to get inside the
hair shaft where they can be ready to react chemically.

## Contents

Aqua
Ethoxydiglycol
Propylene glycol
Isopropyl alcohol
Soytrimonium chloride
Ammonium hydroxide
Oleth-5
Trisodium ethylenediamine
  disuccinate
Tall oil acid
Oleth-2
4-Amino-2-hydroxytoluene
p-Phenylenediamine
Parfum
C11–15 pareth-9
Cocamidopropyl betaine
Resorcinol
p-Aminophenol
C12–15 pareth-3
Citric acid
Erythorbic acid
1-Naphthol
Phenyl methyl pyrazolone
Sodium sulfite
m-Aminophenol
Limonene
Sodium chloride
EDTA
Linalool
1-Hydroxyethyl 4,5-diamino
pyrazole sulfate
Citronellol
Glycerin

## Solvents

These are **aqua,** which is water, **ethoxydiglycol,** which is a gentle solvent with respect to contact with the skin of the scalp, **propylene glycol** and **isopropyl alcohol.**

## Surfactants/Emulsifiers

**Soytrimonium chloride** is a mixture of cationic surfactants. **Oleth-5** and **oleth-2** are non-ionic surfactants, as are **C11–15 pareth-9** and **C12–15 pareth-3.**
**Cocamidopropyl betaine** is a natural surfactant.
**Tall oil acid** acts as an emulsifier and this comes from pine trees.

## Active Agents

**Ammonium hydroxide** has a pH of around 11 and this opens up the cuticles of the hair. Its pH is regulated by the inclusion of a small amount of **citric acid.**
The dye-precursor molecules are **4-amino-2-hydroxytoluene, p-phenylenediamine, resorcinol, p-aminophenol, m-aminophenol, phenyl methyl pyrazolone, 1-naphthol, and 1-hydroxyethyl 4,5-diamino pyrazole sulfate.**
**Trisodium ethylenediamine disuccinate** and **EDTA** are chelating agents.

**Parfum.** Perfume raw material is a complex mixture, some items of which have to be identified on a label because there are people who may be sensitive to them. The ones so identified here are **limonene**, which comes from the peel of oranges; **linalool**, which smells of lavender with a hint of spiciness; and **citronellol**, which is the main fragrance of geraniums and roses.

**Erythorbic acid** is an antioxidant that protects the dye molecules.

**Sodium sulfite** has a role in modifying the dye colour.

**Sodium chloride** is common salt and is used in dyeing. It encourages the dye molecules to move from the solvent on to the hair.

**Glycerin** acts to ensure all the ingredients blend nicely and that water is not lost.

## Step 2: Colour Activating Crème

*The Chemistry*

In this stage there is formation of the dye molecules and that requires the presence of hydrogen peroxide, which has also to be protected and manoeuvred into the hair shaft. The precursor molecules mentioned above will undergo chemical reactions with one another to form the molecules that are responsible for the deep colours. Hydrogen peroxide is a powerful oxidising agent but it is sensitive to metal ions, which can cause it to decompose, hence the need for chelating agents in the mixture to deactivate these ions. It can also be decomposed by certain common bacteria and they too need to be suppressed.

*Contents*

Aqua
Hydrogen peroxide
Acrylates copolymer
Steareth-21
Oleth-2
Oleth-5
PEG-50 hydrogenated palmamide
Acrylates/steareth-20
  methacrylate copolymer
Oleyl alcohol
Etidronic acid
Disodium EDTA
Sodium lauryl sulfate
Simethicone
Tetrasodium pyrophosphate
Phosphoric acid
Sorbitan stearate
PEG-40 stearate
Magnesium chloride
Magnesium nitrate
Cellulose gum
Methylchloroisothiazolinone
Methylisothiazolinone
Potassium sorbate
Sorbic acid

**Aqua** is water and acts as a solvent.

*Surfactants/Emulsifiers*

**Sodium lauryl sulfate** is now only a minor ingredient. Other surfactants are **PEG-50 hydrogenated palmamide, sorbitan stearate**, and **PEG-40 stearate**. In fact the role of these kinds of chemicals is more as emulsifiers to ensure the crème remains consistent. **Acrylates copolymer, steareth-21, oleth-2, oleth-5**, and **acrylates/steareth-20 methacrylate copolymer** also have this dual role.

*Conditioners*

**Oleyl alcohol** is a long-chain alcohol.

**Simethicone** is a mixture of a linear silicone and silicon dioxide and together these suppress foam.

**Hydrogen peroxide** activates the dye precursors prior to their reaction to form the dye.

**Etidronic acid** and **disodium EDTA** act as chelating agents.

**Tetrasodium pyrophosphate** is a water-softening agent.

**Phosphoric acid**, $H_3PO_4$, makes the application fluid acidic.

**Magnesium chloride** and **magnesium nitrate** provide magnesium ions, which some believe have a beneficial effect on the skin.

**Cellulose gum** is a modified form of cellulose that is stable in water and acts as a thickening agent.

**Methylisothiazolinone, methylchloroisothiazolinone**, and **potassium sorbate** are powerful preservatives and will prevent microbes from destroying the hydrogen peroxide.

**Sorbic acid** is an antimicrobial agent.

## Step 3: Colourseal™ Gloss Conditioner

*The Chemistry*

The final step is to close the hair cuticles by lowering the pH. The hair is now treated to ensure it is manageable.

Aqua

Bis-hydroxy/methoxy
  amodimethicone

Stearyl alcohol

Cetyl alcohol

Stearamidopropyl dimethylamine

Glutamic acid

Benzyl alcohol

Parfum

Panthenyl ethyl ether

EDTA

Hexyl cinnamal

*Carthamus tinctorius* seed oil

*Cocos nucifera* oil

Panthenol

Hydrolyzed sweet almond protein

Benzyl salicylate

Magnesium nitrate

Trimethylsiloxysilicate

*Aloe Barbadensis* leaf juice

Sodium benzoate

Methylchloroisothiazolinone

Phenoxyethanol

Magnesium chloride

Methylparaben

Citric acid

Methylisothiazolinone

Propylparaben

Ascorbic acid

Potassium sorbate

Sodium sulfite

**Aqua** is water and acts as a solvent.

**Bis-hydroxy/methoxy amodimethicone** is a modified silicone polymer that acts like a lubricant and allows hair to be combed easily when wet. It also leaves traces on the hair, which have the same effect when the hair is dry.

**Stearyl alcohol** and **cetyl alcohol** are emulsifiers.

**Stearamidopropyl dimethylamine** is attracted to the strands of hair where it acts as an antistatic agent, as does **panthenyl ethyl ether**.

**Glutamic acid** and **citric acid** are weak acids and these lower the pH and this closes the cuticles of the hair.

**Benzyl alcohol** acts as a solvent.

**Parfum.** Perfume raw material is a complex mixture, some items of which have to be identified on a label because there are people who may be sensitive to them. The ones so identified here are **coumarin,** which smells of new-mown hay; and **linalool**, which has a pleasing floral note with a hint of spiciness. **Hexyl cinnamal** smells of camomile.

**Benzyl salicylate** is a fixative; in other words it is there to help the fragrance molecules to blend in with the other ingredients. Of itself, it has almost no odour.

**EDTA** binds to trace metals and prevents them interfering.

*Carthamus tinctorius* **seed oil** is safflower oil, and *Cocos nucifera* **oil** is coconut oil, and these give a sheen to the hair.

**Panthenol** is similar to vitamin B5 and is attracted to the hair and makes it shine.

**Hydrolyzed sweet almond protein** is a natural antistatic hair conditioner extracted from the nuts of the *Prunus dulcis* (sweet almond) tree.

**Magnesium nitrate** and **magnesium chloride** provide magnesium ions, which are supposed to have a beneficial effect on the scalp.

**Trimethylsiloxysilicate** is an anti-foaming agent, emollient, and skin conditioner.

*Aloe barbadensis* **leaf juice** is the liquid more commonly known as aloe vera and is popularly believed to have healing and skin soothing properties.

**Ascorbic acid** is vitamin C and a powerful antioxidant.

**Sodium sulfite** prevents oxygen damage to the other ingredients.

*Antimicrobials*

Some of these are already present in the ingredients that are used to make the product. They are **sodium benzoate, methylchloroisothiazolinone, phenoxyethanol, methylparaben, methylisothiazolinone propylparaben**, and **potassium sorbate**.

There are also hair dyes marketed at women sold under the brand names of Schwarzkoff®, L'Oréal®, and others.

Men may also dye their hair, but products aimed at the male market are usually those in which the coloured molecules are attached to the outside of the hair, such as the following.

**Just for Men™ Black Hair Colour**

*The Chemistry*

This type of product consists of dye molecules designed to colour grey hair. The one described here will dye hair black. There are other versions that colour hair various shades of brown. The product is applied to the hair from the plastic applicator bottle and then worked well in and left for at least five minutes before the hair is shampooed. This product dyes the hair on the outside of each strand.

The dyeing process requires two chemicals: hydrogen peroxide to activate the dye-precursors and a mordant that will connect them to the strands of hair. In the product described here, the mordant is a tin chemical, sodium stannate trihydrate.

## Contents

**Aqua** is water and acts as a solvent.

**Decyl glucoside** is a gentle non-ionic surfactant.

**Ethanolamine** is used as a solvent for the dye precursors.

**Resorcinol, *p*-phenylenediamine, *N*,*N*-bis(2-hydroxyethyl)-*p*-phenylenediamine sulfate, *m*-aminophenol, *p*-aminophenol,** and **2-amino-4-hydroxyethylaminoanisole sulfate** are dye-precursors.

**Carbomer** is the polymer polyacrylate and is added to make the product form a smooth gel.

**Sodium sulfite** prevents oxygen damage to the other ingredients.

**Erythorbic acid** is an antioxidant.

**Parfum.** Perfume raw material is a complex mixture, some items of which have to be identified on a label because there are people who may be sensitive to them. The ones so identified here are ***Chamomilla recutita (Matricaria)* flower extract**, which has the smell of camomile; **coumarin**, which smells of new-mown hay; while **linalool** has a pleasing floral note with a hint of spice.

**Benzyl salicylate** is a fixative; in other words it is there to help the fragrance molecules to blend in with the other ingredients. Of itself, it has almost no odour.

**Trisodium EDTA, disodium EDTA** and **etidronic acid** are chelating agents to stop traces of metal ions and the calcium of hard water from interfering with the dyeing process.

**Caramel** gives the product a pale yellow colour.

**Cinnamido propyltrimonium chloride** is there to protect the dyes from UV damage.

**Laurdimonium hydroxypropyl hydrolyzed wheat protein** is an antistatic hair conditioner.

**Panthenol** is attracted to the hair and makes it shine.

**Tocopheryl acetate** is a form of vitamin E and acts as a protective antioxidant.

*Aloe barbadensis* **leaf juice** is the liquid more commonly known as aloe vera and is popularly believed to have healing and skin soothing properties.

**C11–15 pareth-3** and **C11–15 pareth-9** are non-ionic surfactants.

**Hydrogen peroxide** activates the precursors to react to form the many dye molecules.

**Sodium stannate trihydrate** is the mordant that attaches the dyes to the hair.

## 8.  SHAVING

Humans are very aware of facial hair and this is a natural feature of men. Shaving is commonly used to remove it, and this generally means using a razor whose action can be made easier by providing a lubricating foam, possibly to be followed by a soothing balm.

### Gillette® Fusion Hydra Gel for Sensitive Skin

This contains surfactants, lubricating agents, and emollients (skin conditioners). It also leaves a protective film on the surface of the skin.

*Contents*

Aqua
Palmitic acid
Triethanolamine
Isopentane
Glyceryl oleate
Stearic acid
Parfum
Isobutane
Sorbitol
Glycerin
Hydroxyethylcellulose
PTFE
PEG-90M
Tocopheryl acetate
PEG-23M
Propylene glycol
Glyceryl acrylate/acrylic acid
  copolymer

PVM/MA copolymer
*Aloe barbadensis* leaf juice
Methylparaben
Propylparaben
Alpha-isomethyl ionone
Benzyl alcohol
Benzyl salicylate
Butylphenyl methylpropional
Coumarin,
Eugenol
Hexyl cinnamal
Hydroxyisohexyl 3-cyclohexene
  carboxaldehyde
Limonene
Linalool
CI 42053
CI 42090

**Aqua** is water and acts as a solvent.

**Palmitic acid, triethanolamine**, and **stearic acid** act as emulsifiers.

**Isopentane** and **isobutane** are the propellant gases. These are under pressure in the canister and they force out the content through a nozzle when the pressure is released.

**Parfum.** Perfume raw material is a complex mixture, some items of which have to be identified on a label because there are people who may be sensitive to them. The ones so identified here are **alpha-isomethyl ionone**, which has a 'clean' smell, **butylphenyl methylpropional**, which has a

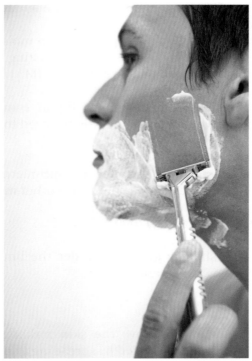

© Shutterstock

fresh outdoor smell; **coumarin** has the smell of new mown hay; **eugenol** has a spicy smell like cloves; **hexyl cinnamal** smells of camomile; **hydroxyisohexyl 3-cyclohexene carbox-aldehyde** smells like lilies and it lingers on the skin; **limo-nene**, which is present in the peel of oranges; and **linalool**, which has a floral note with a hint of spiciness.

**Benzyl salicylate** is a fixative; in other words it is there to help the fragrance molecules to blend in with the other ingredients. Of itself, it has almost no odour.

**Sorbitol** keeps the product moist and viscous.

**Glycerin** is a lubricant and humectant, and makes the skin supple.

**Hydroxyethylcellulose** keeps the shaving gel moist, as do **PEG-90M,** and **propylene glycol**, which also acts as a solvent for the other ingredients. **PEG-23M** acts as a foam booster.

**PTFE** is better known as Teflon. It lubricates the shaving foam.

**Tocopheryl acetate** is able to penetrate the surface of the skin and is an antioxidant.

**Glyceryl acrylate/acrylic acid copolymer** is a moisturiser.

**PVM/MA copolymer** is a copolymer consisting of poly(vinyl methyl ether) PVM and maleic anhydride (MA) and is used as a binder to stabilise the product.

*Aloe barbadensis* **leaf juice** is the liquid more commonly known as aloe vera and is popularly believed to have healing and skin soothing properties.

**Glyceryl oleate** is an emollient.

**Methylparaben** and **propylparaben** are antimicrobials.

**Benzyl alcohol** acts partly as a solvent for other ingredients but is also a preservative.

**CI 42053** is green and **CI 42090** is blue.

There are also shaving gels sold under the brand names of L'Oréal®, Nivea®, Dove®, and others.

### Aftershave Balm

Shaving with a razor can leave the skin rather sensitive and feeling raw. This can be alleviated by spreading a soothing cream on the face and neck.

### Gillette® Sensitive Balm

*Contents*

Aqua
Cetearyl isononanoate
Glycerin
Panthenol
Dimethicone
Carbomer
Xanthan gum
Parfum
DMDM hydantoin
Bisabolol
Myreth-3 myristate
Sodium hydroxide
Acrylates/C10-30 alkyl acrylate
  crosspolymer
Disodium EDTA
Dimethiconol
Tocopheryl acetate
PPG-26 oleate
Limonene
Benzyl salicylate
Linalool
Farnesol
Butylene glycol
Iodopropynyl butylcarbamate
*Aloe barbadensis* leaf juice
Citral
Hydroxyisohexyl 3-cyclohexene
  carboxaldehyde
Pyruvic acid
Alpha-isomethyl ionone
Citronellol
Coumarin
Butylphenyl methylpropional
Geraniol

**Aqua** is water and acts as a solvent.

**Cetearyl isononanoate** is an emollient and skin conditioner.

**Glycerin** keeps the skin moist.

**Panthenol** is related to vitamin B5 and is supposed to have beneficial effects for the skin.

**Dimethicone** is silicone oil, which leaves a layer on the skin that prevents it drying out.

**Carbomer** and **xanthan gum** act as thickening agents.

**Parfum.** Perfume raw material is a complex mixture, some items of which have to be identified on a label because there are people who may be sensitive to them. The ones so identified here are **limonene**, which smells of oranges; **linalool**, which has a floral smell; **citral**, which smells of lemon; **alpha-isomethyl ionone**, which has a clean floral smell; **citronellol**, which smells of geraniums and roses; **coumarin**, which has the smell of new-mown hay; **butyl-phenyl methylpropional**, which has a fresh outdoor smell; **geraniol**, which smells of roses; and **hydroxyisohexyl 3-cyclohexene carboxaldehyde**, which smells of lilies.

**Farnesol** enhances the floral smell of other fragrances.

**Benzyl salicylate** is a fixative; in other words it is there to help the fragrance molecules to blend in with the other ingredients. Of itself, it has almost no odour.

**DMDM hydantoin** is an antimicrobial agent.

**Bisabolol** is a natural complex alcohol believed to have skin healing properties.

**Myreth-3 myristate** is an emollient and skin conditioner.

**Sodium hydroxide** keeps the product slightly alkaline.

**Acrylates/C10-30 alkyl acrylate crosspolymer** forms a protective film on the skin.

**Disodium EDTA** is a chelating agent.

**Dimethiconol** is another type of silicone oil, which also leaves a protective layer on the skin.

**Tocopheryl acetate** is a form of vitamin E and it can penetrate skin where it acts as an antioxidant.

**PPG-26 oleate** is an emollient and skin conditioner.

**Butylene glycol** attracts water molecules and thereby prevents the skin from drying out.

**Iodopropynyl butylcarbamate** is an antimicrobial agent.

*Aloe barbadensis* **leaf juice** is the liquid more commonly known as aloe vera and is popularly believed to have healing and skin soothing properties.

Although **pyruvic acid** is listed in the ingredients it will have been neutralised by the sodium hydroxide and be present as sodium pyruvate. Its presence is no doubt as a stabilising accompaniment to one of the fragrances.

Other aftershave products are available under the brand names of Dove®, Nivea®, L'Oréal® *etc.*

## 9.  HAIR REMOVER

*The Chemistry*

People may wish to remove hair from various parts of their bodies, including their upper lip, armpits, back, or legs. This can be done in a painful way by gluing it to something and pulling it out by its roots, as we will see below with a product designed to remove chest hair. It can also be done in a painless way by using a product that will weaken hair's attachment to the skin. Hair relies on sulfur-to-sulfur bonds for its structure and strength. Break these bonds and the hair becomes detached from its follicles and then can simply be wiped away. A combination of an alkali, which opens the cuticles of the hair, and thioglycolate, which reacts and breaks its sulfur bonds, can do this, and these are the main active agents in the following product.

**Veet**®

*Contents*

| | |
|---|---|
| Aqua | Propylene glycol |
| Urea | Lithium magnesium sodium silicate |
| Potassium thioglycolate | Sodium gluconate |
| Paraffinum liquidum | Acrylates copolymer |
| Cetearyl alcohol | Linalool |
| Calcium hydroxide | Hydrated silica |
| Talc | Propylene glycol dicaprylate/dicaprate |
| Ceteareth-20 | *Nelumbo nucifera* flower extract |
| Glycerin | Phenoxyethanol |
| Sorbitol | Potassium sorbate |
| Potassium hydroxide | Xanthan gum |
| Magnesium trisilicate | CI 77891 |
| Parfum | CI 45380 (3) |

**Aqua** is water and acts as a solvent.

**Urea** helps to rehydrate dry skin.

**Potassium thioglycolate** weakens the hair as described above.

**Paraffinum liquidum** is mineral oil and acts as a viscous solvent.

**Cetearyl alcohol** acts as a thickener and is also as an emollient.

**Calcium hydroxide** and **potassium hydroxide** raise the pH, which makes the hair more porous thereby allowing the other ingredients to penetrate and do their work.

**Talc** is a natural mineral renowned for its softness.

**Ceteareth-20** acts as an emulsifier.

**Glycerin** stops the Veet® from drying out.

**Sorbitol** acts as a lubricant and is non-greasy.

**Magnesium trisilicate** acts as a mild abrasive.

**Parfum.** Perfume raw material is a complex mixture, some items of which have to be identified on a label because there are people who may be sensitive to them. The one so identified here is **linalool,** which smells of lavender and spices.

**Propylene glycol** helps moisturise the skin.

**Lithium magnesium sodium silicate** gives Veet® a slight opalescence.

**Sodium gluconate** acts as chelating agent.

**Acrylates copolymer** is a fluid that blends the other ingredients together.

**Hydrated silica** is a filler.

**Propylene glycol dicaprylate/dicaprate** is an emollient.

*Nelumbo nucifera* **flower extract** comes from the lotus flower, and it contains the chemical quercetin, which has anti-oxidant properties so may act as a preservative.

**Phenoxyethanol** and **potassium sorbate** are preservatives.

**Xanthan gum** acts as a thickening agent.

**CI 77891** is titanium dioxide (white) pigment.

**CI 45380 (3)** is a red dye.

## Nads® For Men Body Waxing Strips

*The Chemistry*

A hairless chest is common among men who appear bare-chested on screen. Unwanted chest hair can be removed by pulling it out by the roots. To be so depilated, they need a product like Nads®. It consists of three components: one contains ingredients to clean and deaden the area to be depilated; then comes the wax strip that, when stuck to the hair, must be strong enough to pull the hair from its roots and yet present no risk to the skin; and finally there is a cleansing wipe containing ingredients to remove any residues, again without risk to the skin.

## Step 1: desensitising wipe

*Contents*

Aqua (water)                          *Piper methysticum* extract
Alcohol                               Sodium hydroxymethylglycinate
Menthol                               Parfum

**Aqua** is water and acts as a solvent.

**Alcohol** removes oil from the hair, so the wax strip will adhere better to it.

**Menthol** makes the skin feel cooler by triggering its cold-sensitive receptors.

***Piper methysticum*** (kava) is a Pacific plant whose extract contains kavalactones, which have a numbing effect on the skin.

**Sodium hydroxymethylglycinate** acts as a preservative.

**Parfum.** Perfume raw material is a complex mixture, some items of which have to be identified on a label because there are people who may be sensitive to them. In this case there are no such fragrances present.

## Step 2: wax strip

*Contents*

| | |
|---|---|
| Triethylene glycol rosinate | *Calendula officinalis* extract |
| Glyceryl rosinate | Guaiazulene |
| Cera microcristalina | Parfum |
| *Mentha piperita* extract | |

**Triethylene glycol rosinate** and **glyceryl rosinate** are the glues that stick to hair strongly enough for it to be pulled from its roots when the adhesive strip is peeled off.

**Cera microcristalina** is a hydrocarbon wax consisting of tiny crystals and these thicken the gum.

***Mentha piperita* extract** is peppermint leaf extract, which will reduce the inflammation of the skin. Its active agent is menthol.

***Calendula officinalis*** extract contains anti-inflammatory agents and antioxidants.

**Guaiazulene** is extracted from camomile oil and is an approved cosmetic colourant.

**Parfum.** Perfume raw material is a complex mixture, some items of which have to be identified on a label because there are people who may be sensitive to them. None needs to be identified here.

## Step 3: soothing & finishing wipes

*Contents*

| | |
|---|---|
| Isopropyl myristate | Tocopheryl acetate |
| Coco-caprylate/caprate | *Prunus amygdalus dulcis* |
| Isohexdecane | *Calendula officinalis* flower oil |

Bisabolol                    Caprylyl glycol
Phenoxyethanol               BHT

**Isopropyl myristate** is a long-chain fatty acid ester that can be absorbed by the skin and helps remove wax that has penetrated the pores.

**Coco-caprylate/caprate** is an emulsifier.

**Isohexdecane** will dissolve and remove wax strip residues.

**Tocopheryl acetate** is a form of vitamin E and it acts as an antioxidant.

*Oils*

*Prunus amygdalus dulcis* is sweet almond oil and an omega-9 fatty acid, *Calendula officinalis* **flower oil** comes from the common marigold, while **bisabolol** is a natural, complex alcohol from the camomile plant. These oils are believed to have skin healing properties.

**Phenoxyethanol** and **caprylyl glycol** are preservatives.

**BHT** or butylated hydroxytoluene is an antimicrobial agent.

## 10.   FAKE TANS

*The Chemistry*

When skin is exposed to sunlight it protects itself against harmful UV rays by darkening. The most effective UV-blocker is a chemical pigment called melanin, which is produced in the upper layers of the skin in cells called melanocytes. The more UV-A is absorbed, the more melanin is formed, and the deeper the colour of the skin becomes. People of African origin are born with naturally high levels of melanin and maintain these levels through life. In fair-skinned individuals, the skin takes time to produce melanin and so it cannot protect against sunburn following sudden exposure.

When you go on holiday, you may wish to expose your body to the sun's rays but may feel embarrassed that your skin is very pale. Some people choose to either take a pre-holiday course of sun-bed treatment (although there are numerous studies raising health concerns relating to the use of sun-beds) or use a fake tan like the one discussed here. The active ingredient in such products is dihydroxyacetone, which combines with proteins in the skin to form molecules that are coloured brown.

Users need to remember that most fake tans do not provide protection against UV rays. What is needed is a sunscreen – see next product.

## St Tropez® Self Tan Bronzing Lotion

What is remarkable about this product is the many fragrances it contains. These appear to be needed to disguise the rather un-usual smell associated with using this product.

*Contents*

Water (aqua/eau)
Dihydroxyacetone
Caprylic/capric triglyceride
Glycerin
*Prunus armeniaca* (apricot) kernel
   oil
Glyceryl stearate
PEG-100 stearate
Cetearyl alcohol
Cetyl alcohol
Sorbitan stearate

Dimethicone
*Aloe barbadensis* leaf juice
*Chamomilla recutita (matricaria)*
   flower extract
Ceteareth-20
Sclerotium gum
Butylene glycol
Potassium sorbate
Phenoxyethanol
Sodium bisulfite
Fragrance (parfum)

Benzyl salicylate
Cinnamyl alcohol
Citronellol
Coumarin
Eugenol
Alpha-isomethyl ionone
Geraniol
Hexyl cinnamal
Hydroxycitronellal
Isoeugenol

Butylphenyl methylpropional
Linalool
Hydroxyisohexyl 3-cyclohexene
  carboxaldehyde
Citric acid
Caramel
CI 15985 (Yellow 6)
CI 16035 (Red 40)
CI 42090 (Blue 1)

**Aqua** is water and acts as a solvent.

**Dihydroxyacetone** reacts with proteins in the outer layer of skin (the stratum corneum) to form chemicals known as melanoidins. These will make the skin turn brown within 30 minutes, giving the appearance of a natural tan. Tanning lotions and creams contain between 2% and 5% dihydroxy-acetone. They have to be mildly acidic, which explains why **citric acid** is present.

**Caprylic/capric triglyceride** is coconut oil and this relaxes and softens the skin, thus improving contact with the dihydroxy-acetone. *Prunus armeniaca* **(apricot) kernel oil** acts similarly.

**Glycerin** acts to ensure all the ingredients blend nicely.

**Glyceryl stearate** acts as a skin conditioning agent as well as being an emollient.

**PEG-100 stearate** is a humectant and moisturiser.

**Cetearyl alcohol** acts as a thickener and boosts foaming (and it also acts as an emollient).

**Cetyl alcohol** is used both as an emollient and to make the liquid look opaque.

**Sorbitan stearate** acts as an emulsifier

**Dimethicone** is a colourless silicone oil that leaves behind a protective, water-repelling silicone layer on the skin.

*Aloe Barbadensis* **leaf juice** is the liquid more commonly known as aloe vera, and is popularly believed to have healing and skin soothing properties.

**Ceteareth-20** acts as a non-ionic surfactant.

**Sclerotium gum** is a carbohydrate derived from fungi and it readily forms a gel (jelly) in water. It thickens and emulsifies the product and also helps make the skin soft and smooth.

**Butylene glycol** is also a humectant naturally present in skin, and prevents loss of water from the skin.

**Potassium sorbate, phenoxyethanol,** and **sodium bisulfite** are preservatives.

**Fragrance (parfum).** Perfume raw material is a complex mixture, some items of which have to be identified on a label because there are people who may be sensitive to them. The ones so identified here are ***Chamomilla recutita* (*Matricaria*) flower extract,** which has the smell of camomile, as does **hexyl cinnamal. Cinnamyl alcohol** smells of hyacinth, **citronellol** is the main fragrance of geraniums and roses, **coumarin** has the scent of new-mown hay, **eugenol** has a spicy smell like cloves, **alpha-isomethyl ionone** has a 'clean' smell, **geraniol** smells of roses, **hydroxycitronellal** smells of lime, **isoeugenol** smells of vanilla, **butylphenyl methylpropional** has a fresh outdoor smell, **linalool** has a pleasing floral note with a hint of spiciness, and **hydroxyisohexyl 3-cyclohexene carboxaldehyde** smells of lilies.

**Benzyl salicylate** is a fixative; in other words it is there to help the fragrance molecules to blend in with the other ingredients.

**Caramel** is polymerised sugar, and this gives the product a pale yellow colour. In the ingredient list for this product the colours of the dyes are spelled out, namely **CI 15985 (yellow 6), CI 16035 (red 40)** and **CI 42090 (blue 1).**

## 11.   SUNSCREEN

*The Chemistry*

There are two kinds of UV rays that we need to protect ourselves against: UV-A rays that cause the skin to darken, although this may be what we want to happen; and UV-B rays that leave your skin red and raw, which we don't want to happen. Sunscreens can protect us against too much of either.

A good sunscreen should include titanium dioxide pigment to protect against UV-A plus an appropriate molecule to protect against UV-B. The titanium dioxide particles need to be nano-sized, which means they are around 50 nanometres in size and so are invisible to the naked eye.

The sunscreen should be easy to apply and give a continuous film. It should not be sticky, nor washed off when swimming, although it should be easily removed when using a shower gel. Nor should it harbour germs, and it should smell nice.

The key point to look for when choosing a sunscreen is its sun protection factor (SPF). This is an indication of the extent to which it blocks out UV rays. Factor 2 cuts UV by half, in other words to 50%, a factor 4 reduces it to a quarter (25%), while factor 20 reduces it to 5%. It is possible to have factor 50, which

© Shutterstock

would block 98% of the UV, but some exposure of the skin to UV is beneficial because it promotes the formation of vitamin D. However, some people are more prone to sunburn than others and which sunscreen you choose will also depend on age (very young children may need extra protection) and on the length of time you will be exposed to the sun.

In June 2014, the monthly magazine of the Consumers Association, *Which?*, reported on sunscreens and named seven brands as 'best buys.' They were Avon®, Boots Soltan®, Calypso®, Garnier Ambre Solaire®, M&S® Formula, Superdrug Solait®, and Nivea®.

## Nivea Sun® Protect & Refresh Lotion SPF 20

*Contents*

Aqua
Alcohol denat
Homosalate
Butyl methoxydibenzoylmethane
Ethylhexyl salicylate
Octocrylene,
Tapioca starch
Isopropyl palmitate
Titanium dioxide (nano)
Silica dimethyl silylate
Glycerin
Menthol
Tocopheryl acetate
Sodium acrylates/C10–30 alkyl
 acrylate crosspolymer
Hydroxyethylcellulose
Carbomer
Trimethoxycaprylylsilane
Trisodium EDTA
Phenoxyethanol
Linalool
Limonene
Butylphenyl methylpropional
Benzyl alcohol
Alpha-isomethyl ionone
Citronellol
Eugenol
Coumarin
Parfum

*UV Blockers*

**Homosalate, butyl methoxydibenzoylmethane, ethylhexyl salicylate** and **octocrylene** filter out mainly UV-B rays, although octocrylene also absorbs UV-A rays. Homosalate has a second benefit in that it makes the sunscreen waterproof.

**Titanium dioxide (nano)** is present as particles that are so small individually that they remain invisible yet absorb and reflect the UV-A rays.

**Aqua** is water and acts as a solvent.

**Alcohol denat** is denatured alcohol, which contains a bitter-tasting chemical to deter its theft and misuse. (By denaturing the alcohol, the company also avoids paying excise duty.) Here it acts, along with water, as the solvent for the various ingredients.

**Tapioca starch** and **isopropyl palmitate** are used as thickeners and emulsifying agents.

**Silica dimethyl silylate** and **sodium acrylates/C10–30 alkyl acrylate crosspolymer** form a film of lotion on the skin.

**Glycerin** acts as a thickening agent and humectant.

**Menthol** has a cooling effect because it can trigger cold-sensitive receptors when it comes in contact with the skin.

**Tocopheryl acetate** is vitamin E and acts as an antioxidant.

**Hydroxyethylcellulose** is a viscosity controller.

**Carbomer** is another name for polyacrylate and is added to form a smooth gel.

**Trimethoxycaprylylsilane** is a silicone oil which, together with the titanium dioxide, forms a film that spreads evenly and is stable.

**Trisodium EDTA** is a chelating agent to stop trace metal ions from interfering with the action of the other ingredients.

**Phenoxyethanol** and **benzyl alcohol** are preservatives.

**Parfum.** Perfume raw material is a complex mixture, some items of which have to be identified on a label because there are people who may be sensitive to them. The ones so identified here are: **linalool**, which has a floral spicy smell; **limonene**, which smells of oranges; **butylphenyl methylpropional** and **alpha-isomethyl ionone**, which have a clean outdoor smell; **citronellol**, which smells of geraniums and roses; **eugenol**, which has a spicy smell like cloves; and **coumarin**, which smells of new-mown hay.

## 12.  DEODORANT AND ANTIPERSPIRANT

*The Chemistry*

Bacteria in our armpits feed on the sweat that our pores release and their action can cause offensive odours, the most recognisable one being the chemical *trans*-3-methyl-2-hexenoic acid. This isn't the only chemical that is produced, and the sweat of men can be particularly rich in chemicals and much stronger smelling than female underarm sweat.

We can mask the smell with perfumes, deodorise it, or prevent it being formed. Some products are designed to kill the offending bacteria with antibacterial agents while others block pores and so prevent them accessing the sweat. (This last kind of product also prevents underarm dampness, which can be embarrassing and may even stain clothing.) These agents will come dissolved in a solvent, a gel, or a soft wax.

Blocking the pores is done with plugs of insoluble aluminium or zirconium hydroxide.

## Dove® Deodorant

*Contents*

| | |
|---|---|
| Aqua (water) | Alpha-isomethyl ionone |
| Aluminium chlorohydrate | Benzoyl alcohol |
| Glycerin | Benzyl salicylate |
| *Helianthus annuus* seed oil | Butylphenyl methylpropional |
| Steareth-2 | Citronellol |
| Parfum | Geraniol |
| Steareth-20 | Hexyl cinnamal |
| Citric acid | Limonene |
| Potassium lactate | Linalool |

**Aqua** is water and acts as a solvent.

**Aluminium chlorohydrate** reacts with water in the pores to form insoluble plugs of aluminium hydroxide that blocks the pores.

**Glycerin** is a humectant and is there to keep the deodorant moist.

***Helianthus annuus* seed oil** is sunflower oil and, as an emollient, soothes and softens the skin.

**Steareth-2** and **steareth-20** are emulsifying agents to blend all the other ingredients into a homogeneous mixture. **Potassium lactate** is another emulsifier.

**Parfum.** Perfume raw material is a complex mixture, some items of which have to be identified on a label because there are people who may be sensitive to them. The ones so identified here are **alpha-isomethyl ionone**, which has what is described as a 'clean' smell; **butylphenyl methylpropional**, which has a fresh outdoor smell; **citronellol**, which smells of geraniums and roses; **geraniol**, which smells of roses; **hexyl cinnamal**, which smells of camomile; **limonene**, which smells of oranges and lemons; and **linalool**, which also has a floral smell with a hint of spice.

**Benzyl** alcohol is a preservative.

**Benzyl salicylate** is a fixative; in other words it is there to help the fragrance molecules to blend in with the other ingredients. Of itself, it has almost no odour.

**Citric acid** is a skin conditioning agent.

*More Chemistry*

When aluminium was suspected of causing Alzheimer's disease, there were attempts to reduce human exposure to this metal of which there were many uses, including underarm deodorants. Another metal that behaves very like aluminium in that it forms an insoluble hydroxide is zirconium, and this was introduced to replace aluminium in underarm deodorants. Although it was more expensive, it appeared to work better. Even when, in the 1990s, the alarms about aluminium were shown to be false, this type of underarm deodorant continued to be used along with aluminium compounds, which have been reinstated. The following product is of this type.

## Sure® Cool Blue Fast Drying

*Contents*

Aqua
Aluminium zirconium
  pentachlorohydrate
Parfum
Hydroxypropylcellulose
Silica
Amyl cinnamal
Benzyl alcohol

Benzyl benzoate
Citronellol
Coumarin
Geraniol
Hexyl cinnamal
Hydroxycitronellal
Limonene
Linalool

**Aqua** is water and acts as a solvent.

**Aluminium zirconium pentachlorohydrate** provides both aluminium and zirconium ions, which can block the sweat glands by forming the insoluble hydroxides.

**Parfum** is a mixture of fragrances, some natural, some artificial, and it may contain items that some individuals are sensitive to. These are identified as **amyl cinnamal**, which is a fragrance similar to jasmine; **citronellol**, which is the main fragrance of geraniums and roses; **coumarin**, which has the scent of new-mown hay; **geraniol**, which smells of roses; **hexyl cinnamal**, which smells of camomile; **hydroxycitronellal**, which smells of lime; **limonene**, which smells of oranges; and **linalool**, which has a pleasing floral note with a hint of spiciness.

**Hydroxypropylcellulose** is a polymer used as a thickening agent and lubricant.

**Silica** is silicon dioxide and is a filler.

**Benzyl alcohol** is a fixative for the fragrances along with **benzyl benzoate**. They not only protect the deodorant against microbial contamination but also prevent them from breeding in the armpits and so forming unpleasant odour molecules.

Other deodorants and antiperspirants are by L'Oréal®, Sure®, Lynx®, and Right Guard® *etc.*

## 13.  SHOWER CLEANERS

*The Chemistry*

The walls of a shower eventually become stained by limescale, dried water drops, and the residues from the shower soaps and gels we use when cleaning ourselves. These can be removed with acids and surfactants. The shower can also be sanitised by including disinfectants in the cleaner.

### Mr Muscle® Shower Shine

*Contents*

| | |
|---|---|
| Aqua | Dipropylene glycol |
| Decyl glucoside | Parfum |
| Lauryl polyglucose | Benzisothiazolinone |
| Citric acid | Sodium hydroxide |
| Sodium citrate | |

**Aqua** is water and acts as a solvent.

**Decyl glucoside** is a non-ionic surfactant and gentle enough not to affect the skin of anyone standing in the shower as they clean it.

**Lauryl polyglucose** is another gentle non-ionic surfactant that is also used in cosmetics.

**Citric acid** is used to ensure the cleaner is acidic and as such will remove limescale.

**Sodium citrate** acts both as an emulsifier and as an acidity regulator, keeping the pH stable.

**Dipropylene glycol** is a solvent and will help remove greasy marks.

**Parfum** is a mixture of fragrances, some natural, some artificial, and it may contain items that some individuals are sensitive to. Those used here do not come into this category.

**Benzisothiazolinone** is designed to ensure the shower cleaner does not harbour germs.

[**Sodium hydroxide** seems to have been listed because it is present in one of the ingredients. However, in the presence of citric acid it will have been converted to sodium citrate.]

Other shower cleaners are available, for example those under the brand names Dettol® and Jeyes®. However, if limescale is a serious problem on tiles, sinks, baths, taps, and the toilet, then you may need a stronger acid and the following contains two, along with surfactants.

## Scale Away®

*Contents*

| | |
|---|---|
| Water | Methyl cedryl ketone |
| Phosphoric acid | Nonanol-1 |
| Alcohols, C9–11, ethoxylated | α-Hexylcinnamaldehyde |
| Gluconic acid | Non-ionic surfactants |
| *N,N*-dimethyltetradecylamine | 3,7-dimethyloct-6-enenitrile |
|   *N*-oxide | Coumarin |
| Parfum | Eugenol |
| 2-(4-*tert*-butylbenzyl)- | 1-(2,6,6-trimethyl-2-cyclohexen-1- |
|   proprionaldehyde |   yl)-2-buten-1-one |
| 4-*tert*-butylcyclohexyl acetate | Phthalocyanine dye preparation |
| Allylamylglycolate | |

**Water** acts as a solvent.

**Phosphoric acid** is the acid that attacks the limescale. It does not affect any metals such as chrome that are part of the shower.

**Alcohols, C9–11, ethoxylated** are non-ionic surfactants.

**Gluconic acid** is an acid that forms a soluble salt with calcium.

***N,N*-dimethyltetradecylamine *N*-oxide** is used to ensure that the foam remains in contact with the surface to be cleaned.

**Parfum.** Perfume raw material is a complex mixture, some items of which have to be identified on a label because there are people who may be sensitive to them. The ones so identified here are **2-(4-*tert*-butylbenzyl)proprionaldehyde**, which has a floral smell like lily of the valley; **4-*tert*-butylcyclohexyl acetate**, which has a woody floral note; **allylamylglycolate**, which has a fruity pineapple note; **methyl cedryl ketone**, which has a woody amber odour; **nonanol-1**, which has a citrus odour, **α-hexylcinnamaldehyde**, which is like jasmine; **3,7-dimethyloct-6-enenitrile**, which is also known as citronellyl nitrile and it has a citrus odour; **coumarin**, which has the scent of new-mown hay; **eugenol** has a spicy smell like cloves; **1-(2,6,6-trimethyl-2-cyclohexen-1-yl)-2-buten-1-one**, which is better known as alpha-damascone, and it has a floral odour with woody undertones.

**Non-ionic surfactants** are not specified.

**Phthalocyanine dye preparation** is a combination of blue and green dyes.

CHAPTER 4

# The Desk

© Shutterstock

Working from home, as many people do, means working *at* home for part of the week, and requires a desk at least, and ideally a study. There are several items that might be found on a desk, or in its drawers, and some of these involve chemistry.

Chemistry at Home: Exploring the Ingredients in Everyday Products
By John Emsley
© John Emsley, 2015
Published by the Royal Society of Chemistry, www.rsc.org

1. Sticky stuff (Pritt®, Blu-Tack®)
2. Correction fluid (Tipp-Ex® Rapid, Tipp-Ex® Ecolutions Correction Fluid)
3. Writing implements (Pencils, Ballpoint Pens, Gel Pens)
4. Messages (Post-it® Notes)
5. Screen cleaner (3M® Screen & Keyboard Cleaner)
6. Low energy light bulbs (Compact Fluorescent Lamp, LEDs)
7. Chewing gum (Wrigley's Extra™ Peppermint Chewing Gum)

## 1. STICKY STUFF

There are times when we need to stick one thing to another, sometimes permanently, sometimes temporarily. **Pritt**®**glue** is ideal for sticking paper to paper, and **Blu-Tack**® is best when we want to be able to stick something to a surface and remove it later without it leaving a mark. (In Chapter 6 we will look at a stronger adhesive, and at wallpaper paste.)

## Pritt® Glue

*The Chemistry*

A Pritt® stick spreads glue just where it is needed and is ideal for sticking paper and card. Its lipstick-like container makes it clean and easy to use. Originally the glue was polyvinylpyrrolidone but the makers, Henkel of Germany, have replaced this with a glue based on a starch.

*Contents*

| | |
|---|---|
| Starch ether | Hydrogen peroxide |
| Caprolactam | Sodium hydroxide |

**Starch ether** is starch that has been treated with propylene oxide, which converts this natural carbohydrate polymer into a more stable material. It accounts for 95% of the stick. The starch comes from plants like tapioca.

**Caprolactam** accounts for a few per cent, and this interacts with the starch to make it a more solid paste, which it does by forming temporary cross-links between polymer strands.

**Hydrogen peroxide** (less than 1%) is a gentle but powerful bleaching agent.

**Sodium hydroxide** (less than 1%) ensures the glue does not become acidic, which would depolymerise the starch.

Other common glues for sticking paper and card are made by UHU®, Baker Ross®, *etc.*

## Blu-Tack®

*The Chemistry*

There are times when the bond between two things should not be permanent, yet should be enduring enough to hold them together for as long as we wish. Sticking a photograph or a poster to a wall is a typical use for such a product, of which the best known is Blu-Tack®. This is a kind of plastic that acts as a semi-fluid when under pressure.

Blu-Tack® was invented by Alan Holloway in 1970 at a company called Ralli Bondite, which made sealants of the type used in the building trade. A traditional sealant is putty, which is a mixture of linseed oil and calcium carbonate, but it has the disadvantage that the oil can become oxidised and harden over time. Linseed oil is obtained from flax seeds and is prone to this because it is mainly made up of polyunsaturated fatty acids. It appears to have been experiments with other kinds of oil that led to the material that was eventually to become Blu-Tack®.

*Contents*

Mineral filler (80%)          Synthetic rubber (10%)
Mineral oil (10%)             Pigment (<1%)
Xylene (<1%)

**Mineral filler (80%)** is probably calcium carbonate or china clay, which is kaolinite and is an aluminium silicate. Calcium carbonate is the traditional mineral in glazing putty. This is mixed with linseed oil to form such putty.

**Mineral oil (10%)** is a colourless mixture of long chain hydrocarbons extracted from oil and consists of saturated and aromatic components, with the saturated part accounting for almost all of it. The presence of **xylene** (<1%) in the oil has to be reported, by law because this chemical is classified as toxic, although is it in no way dangerous at the level used in Blu-Tack®.

**Synthetic rubber (10%)** is styrene-butadiene rubber (SBR) and isobutylene-isoprene polymer. It is this that gives the Blu-Tack® its mouldability.

**Pigment (<1%).** On its own, Blu-Tack® would be white and it was thought that this might encourage children to eat it; so it had to be given a colour that would not attract them but would act as a warning. Cobalt blue has the shade of blue that is characteristic of Blu-Tack®. Ultramarine would also provide this kind of colour.

## 2.  CORRECTION FLUID

*The Chemistry*

When you make a mistake on something that cannot be replaced easily, such as an original document, you can cover it with a thin film of white paint that quickly dries, and it can then be written over using a ballpoint pen. Such correction fluid was once a part of typing, since it could be typed over. Today it is less used, but nevertheless most desks are likely to have a bottle.

Originally the solvent in correction fluid was 1,1,1-trichloroethane, which is now banned because of its ability to affect the Earth's ozone-layer. Today the solvent is a mixture of medium-sized hydrocarbon chains typical of those in paraffin although these have been treated to ensure that they contain no aromatic hydrocarbons.

The most popular kind of correction fluid is Tipp-Ex® and indeed it became so popular as to enter the language as the verb 'to tippex,' meaning to remove something you want to change or hide.

### Tipp-Ex® Rapid

*Contents*

Aliphatic isoparaffinic hydrocarbon
Aliphatic hydrocarbon
Titanium dioxide

**Aliphatic isoparaffinic hydrocarbon** is a mixture consisting mainly of C12–C15 branched hydrocarbons, some of which are cyclic molecules.

**Aliphatic hydrocarbon** is the same except the chains are linear. Together the solvents have boiling points around 100 °C and they evaporate fairly quickly.

**Titanium dioxide** is a white pigment with remarkable covering power.

### Tipp-Ex® Ecolutions Correction Fluid

Tipp-Ex® Ecolutions Correction Fluid is based on chemicals that are seen as more environmentally friendly, hence the new name.

*Contents*

The solvent here is **ethanediol,** which is also known as ethylene glycol or just glycol. This is a non-hydrocarbon solvent
obtained from renewable resources.

**Titanium   dioxide** is   a   white   pigment   with   remarkable
covering power.

**Ammonia solution** keeps the product alkaline so that it does
not chemically react with paper.

**Limestone** is calcium carbonate (chalk) and is another white
pigment.

### 3. WRITING IMPLEMENTS

## Pencils

*The Chemistry*

In fact there is very little chemistry involved.

Pencils consist of graphite, as the marker, inside a wooden cylinder for which cedar wood is preferred. Although the marker is referred to as the 'lead' of the pencil, this misnomer came about because the large deposit of graphite discovered in Cumbria in the north of England in the 1500s was mistakenly thought to be a lead ore and given the name black lead. The lead of pencils is a mixture of graphite and clay, and the more clay, the harder the lead. Pencils are graded from H (for hardness) to B (for blackness), so there were HH for the pencils used by draftsmen because they keep a fine tip, and BB for those used by artists because this can easily be rubbed out. (HB is the type in general use.)

© Shutterstock

At the end of the pencil is often a rubber (eraser), or you can buy specially shaped rubbers to attach to the end of pencils. These are made of polymers, such as isoprene-isobutylene, known as butyl rubber, or more likely PVC, to which an abrasive like pumice (powdered volcanic rock) has been added.

## Pens

Almost no one now uses a fountain pen and ink for writing. Indeed very little letter writing as such is done any more although we still need pens, if only to sign our name. There are basically two kinds of modern pen: the ballpoint pen, originally called the biro after its Hungarian inventor László Jozsef Biró; and the gel pen, which writes more like traditional pen and ink. So what chemicals do they require?

## Ballpoint Pens

### The Chemistry

It was Biro László, the Hungarian journalist, who engineered the ballpoint of the pen, while his brother György, who was a chemist, developed the ink that made it workable. This needed to be viscous, yet still flow past the ball at the tip of the pen and on to the paper. László also wanted a pen whose ink would dry quickly. Clearly such a pen could not use a water-based ink and so oil-based inks were originally used. Today the solvents are ethylene glycol or propylene glycol. There also needs to be a polymer like alkyd resin to bind the ingredients together. A small amount of surfactant will lower the surface tension and will keep all the ingredients blended.

Typical colouring agents are carbon black, which is a kind of soot; triarylmethane-based ones, which can be a variety of shades from green to violet; and eosin, which is red. Metallic inks use aluminium powder for a silver effect and copper–zinc powder for a gold effect.

*Removing ballpoint ink stains is best done using 2-propanol (also known as isopropyl alcohol or rubbing alcohol) and a paper towel.*

## Gel Pens

Today we have gel pens, which also have a ball tip, but whose ink flows more like the traditional ink although drying more quickly. The ink in a typical gel pen is as follows:

*Contents*

Water
Propylene glycol
Ethylene glycol
Colouring agents

Glycerine
Resin
Polyacrylate

**Water, propylene glycol** and **ethylene glycol** are solvents. They will allow the ink to flow freely and dry evenly.

**Colouring agents** are various dyes, although for metallic-looking inks aluminium powder is used.

**Glycerine** acts to thicken the gel.

**Resin** helps ensure the ink sticks to the surface being written on.

**Polyacrylate** acts to form a firm film on the surface of the ink.

## 4.  MESSAGES

Sometimes you need to write a reminder on a piece of paper that you can stick in a prominent position but which can easily be removed when it is no longer required. The product that meets these requirements is Post-it®.

### Post-it® Notes

*The Chemistry*

These were invented 35 years ago by chemical engineer Art Fry and chemist Spencer Silver. They worked for the US company 3M®, formerly called the Minnesota Mining & Manufacturing company. Spencer's research project was aimed at finding a super-strong adhesive, but one of his new glues proved to have only poor adhesion. However, that was just what was needed to hold temporary notices.

To begin with, all Post-it® notes were yellow in colour and this came about because the paper that Art Fry first used was taken from the office next door to his lab where there was a pile of scrap paper, all of which was yellow. Today of course these notes come in all shapes, sizes, and colours and other companies now make them, although the term 'Post-it®' is still a registered brand name of 3M®.

So what's special about Post-it® note glue? It is a cross-linked acrylate copolymer that forms itself as microspheres and doesn't spread as a tacky film over the surface it comes in contact with. This makes it weak as an adhesive, which is just what is needed, but it works as a temporary fix, and it doesn't leave any of the adhesive behind when the note is removed.

## 5.  SCREEN CLEANER

*The Chemistry*

There are various products for cleaning computer screens and keyboards, some of which contain an anti-static ingredient while others claim to kill germs. They need to include a surfactant. Ideally, after cleaning, the product should leave a protective film on the surface that will neutralise static electricity so that the keyboard and screen will not attract dust. The following is a typical product.

### 3M® Screen and Keyboard Cleaner

*Contents*

| | |
|---|---|
| Water | Sulfonic acid derivative |
| Acrylic polymer | Triethanolamine |
| Sodium lauryl sulfate | Sodium hydroxide |

**Water** is the solvent and it makes up 90% of the cleaner.

**Acrylic polymer** is sodium polyacrylate, which leaves a film on the surface of the screen. It acts as the anti-static agent by virtue of it being ionic and therefore able to deal with a

© Shutterstock

build-up of electrons, which cause the static and so attract dust.

**Sodium lauryl sulfate** is an anionic surfactant.

**Sulfonic acid derivative** is not identified as such but is likely to be one of the sodium dodecylbenzene sulfonates, which may also act both as a charge disperser and as a surfactant.

**Triethanolamine** acts as a non-ionic surfactant.

**Sodium hydroxide** accounts for less than 0.1% and is there to ensure the spray is not acidic.

## 6. LOW ENERGY LIGHT BULBS

When we are working we appreciate light that is focussed and we may use a reading lamp. We also need illumination in the form of tiny lights that indicate something is switched on, such as our laptop, Wi-Fi, or printer. Ideally, all these sources of light should use as little energy as possible.

*The Chemistry*

Electricity can produce light and in various ways. One, very inefficient way, was to heat a wire until it glowed white hot but then more power was dissipated as heat than as light. A much better way was the fluorescent lamp in which an electric discharge was passed through an inert gas. Even better is to use the light emitted by diodes when they are activated.

### Compact Fluorescent Lamp

In this there is argon, along with a tiny amount of mercury. When an electric discharge passes through the gas it excites atoms of this metal and they emit UV light of wavelength 254 nm. This invisible light is then made visible by activating the phosphor coating on the inside surface of the lamp, which converts it to wavelengths in the visible range 400 to 700 nm. The phosphors are compounds of lanthanide metals, the so-called rare earth metals, which emit light in the blue, green, and red parts of the spectrum – chiefly in the 450, 550, and 610 nm wavelengths respectively, which together we perceive as white. The blue emission band comes from atoms of the metal europium, the green from a mixture of lanthanum, cerium, and terbium, and the red from a mixture of yttrium and europium.

### LEDs (Light-Emitting Diodes)

LEDs based on aluminium gallium arsenide first appeared in the 1960s when they were the red figures in numerical displays. Other colours came later, and eventually in the 1990s a combination of red, green, and blue LEDs made white lights possible. Some LEDs just rely on the blue light of gallium nitride in combination with a phosphor that converts its light to white light.

The phosphor is yttrium aluminium garnet (YAG). Light-emitting diodes are tiny and use very little energy. However, they do need some rather unusual elements, such as gallium, arsenic, indium, yttrium, and selenium.

## 7.   CHEWING GUM

*The Chemistry*

Chewing gum can help you to concentrate while satisfying the desire to snack without taking in calories.

Originally, chewing gum was based on natural chicle, made from the latex of the sapodilla tree of Central America. Modern chewing gum is made from synthetic elastic polymers (elastomers). Those approved for use include styrene-butadiene rubber (SBR), iso-butylene-isoprene copolymer, polyethylene, and polyisobutylene.

The essential ingredients in all chewing gums are the gum base, sweeteners, flavours, emulsifiers, humectants, and preservatives.

Polymers become softer when they absorb oil and the same thing happens to gum bases when waxes are added; these make the gum more workable because they act as lubrication between the strands of polymer. Natural waxes include carnauba wax and beeswax. Petroleum waxes are better in that they give a longer shelf life to the product.

Other important ingredients in chewing gum are emulsifiers, such as lecithin and glycerylmonostearate, which are needed to soften it and enable the various components to blend together to form an homogeneous mixture. Humectants are included to prevent the gum drying out and going hard. The best humectant is glycerol. Emulsifiers and humectants together constitute less than 1% of the product.

Various sweeteners are added to chewing gums, such as xylitol, aspartame, and acesulfam, none of which causes tooth decay. (The earlier gum used sugar as the sweetener.)

Flavours account for around 1% of a chewing gum and the most popular are spearmint, which relies on the chemical carvone, and peppermint, which is mainly menthol. Chewing gum also needs antioxidants, such as BHT (short for butylated hydroxytoluene) or BHA (short for butylated hydroxyanisole).

## Wrigley's Extra® Peppermint Chewing Gum

*Contents*

| | |
|---|---|
| Sorbitol | Aspartame |
| Mannitol | Acesulfame K |
| Maltitol | Gum base |

Carnauba wax                    Peppermint
Lecithin                        E171
Gum arabic                      BHA
Glycerol (non-animal)

*The coating*

Sorbitol is a natural sweet-tasting alternative to sugar that does
   not cause tooth decay. Although it occurs naturally in some
   fruits, it is manufactured from glucose.

Mannitol has the same chemical formula as sorbitol but a
   slightly different molecular structure. It has similar sweetness to sorbitol.

Maltitol is made from starch and is almost as sweet as sugar,
   but unlike sugar, it does not cause tooth decay.

Gum arabic is the sap of the acacia tree and is a mixture of
   carbohydrates and proteins. It is added to lots of foods as a
   thickening agent.

E171 is titanium dioxide, which gives the gum coating its
   bright whiteness.

*The gum*

Aspartame and acesulfame K are artificial sweeteners.

Gum base is the synthetic styrene-butadiene rubber (SBR) and
   isobutylene-isoprene polymer. It is this that gives the gum
   its chew.

Carnauba wax is there to lubricate the gum.

Lecithin is the emulsifier and lubricant that not only softens
   the gum but helps all the components blend to an homogeneous mixture. Lecithin is extracted from animal and
   plant tissues but that in Extra gum comes only from
   soybeans.

Glycerol acts as a humectant and it stops the gum drying out
   and becoming hard.

For its flavour, peppermint relies on menthol, which can
   trigger the cold-sensitive receptors on the tongue, producing
   a feeling of freshness.

BHA (butylated hydroxyanisole) is an antioxidant.

CHAPTER 5

# The Toilet

© Shutterstock

Chemistry at Home: Exploring the Ingredients in Everyday Products
By John Emsley
© John Emsley, 2015
Published by the Royal Society of Chemistry, www.rsc.org

This may be the smallest room in the house but it is one of the most important. We want it to be clean and fresh when we enter and we should always leave it the same way for the next person. We also need to wash our hands when we have finished using it. Chemistry has a part to play in cleanliness, as it does in various other products we may use in the privacy of the toilet and these are described here. They are as follows:

1. Toilet cleaner (Harpic® Power Plus)
2. Toilet wipes (Andrex® Washlets™)
3. Handwash (Carex® Moisture Plus, Carex® Antibacterial Handwash with Silver, Palmolive® Ayurituel Joyous Liquid Handwash)
4. Hand cream (Nivea® Creme, Neutrogena® Norwegian Formula® Unscented)
5. Absorbent products (Pampers®, Always® Panty Liners)
6. Suppositories (Anusol®)
7. Pregnancy test kit (First Response® Pregnancy Test Kit)
8. Air fresheners (Air Wick®, Oust®, Febreze® Air Effects)

# 1. TOILET CLEANER

*The Chemistry*

A toilet cleaner generally contains an acid (to remove limescale), a surfactant (to remove faecal traces), a disinfectant (to kill germs), a thickening agent (so that it clings to the sides of the toilet bowl), and a fragrance.

The traditional way of sanitising a toilet was with bleach and this is still the most effective antimicrobial chemical, although some consumers have been sensitised to its use by alarms in the media. In any case, it should not be used in conjunction with an acid descaler because then it will react to form chlorine gas, which is very unpleasant. Some toilet cleaners now rely on hydrogen peroxide to carry out the disinfecting.

The following product uses neither and relies on its high level of hydrochloric acid, which is effective at removing limescale as well as acting as an antimicrobial agent.

## Harpic® Power Plus

*Contents*

| | |
|---|---|
| Aqua | Alkylethoxylate C10-16, 10EO |
| Hydrochloric acid | Alkylethoxylate C9-11, 5.5EO |
| PEG-2 tallow amine | Parfum |
| Tallowtrimethylammonium chloride | Colorant |

**Aqua** is water and the solvent.

**Hydrochloric acid** is 9% strength and so will remove limescale. Limescale is insoluble calcium carbonate which reacts with the acid, releasing the carbonate as $CO_2$ and the calcium as water-soluble calcium chloride so it gets flushed away. Acid of this strength will also kill germs effectively.

**PEG-2 hydrogenated tallow** amine is both a surfactant and an emulsifier and is needed to keep all the other ingredients well mixed.

**Tallowtrimethylammonium chloride** acts as a thickener.

**Alkylethoxylate C10-16, 10EO** and **alkylethoxylate C9-11, 5.5EO** are non-ionic surfactants.

The **parfum** ingredients are not specified, which means that
none of the fragrance ingredients is of the kind that some
people are sensitive to.

**Colorant** is not specified.

Other toilet cleaners are available, such as those under the
brand names Domestos®, Duck®, Cillit Bang®, Mr Muscle®,
and Ecover®.

## 2.  TOILET WIPES

*The Chemistry*

These are to be used as a final wet wipe to clean and sanitise the skin, and for that purpose they include surfactants and germ-killing agents. They are like toilet paper in that they are intended to disintegrate and disperse when flushed away (whereas paper handkerchiefs and paper towels are meant to remain strong when wet and should never be disposed of down a toilet). Toddler wipes should also not be flush down the toilet.

### Andrex® Washlets™

These can be used to wipe yourself after you have used the toilet.

*Contents*

| | |
|---|---|
| Aqua | Lauryl glucoside |
| Sodium chloride | Parfum |
| Sodium benzoate | Methylisothiazolinone |
| Polysorbate 20 | Aloe barbadensis extract |
| Sodium lauryl glucose | Propylene glycol |
| carboxylate | Tocopheryl scetate. |
| Malic acid | |

**Aqua** is water and acts as the solvent.

**Sodium chloride** is common salt and this ensures the wipes remain moist and don't dry out even when the pack has been opened.

**Polysorbate 20** ensures the ingredients of the wetting solution remain smoothly blended.

*Aloe barbadensis* **extract,** which is more commonly known as aloe vera, is supposed to have a soothing effect and to promote healing.

**Propylene glycol** is a solvent with roughly the same dissolving properties as alcohol.

**Tocopheryl acetate** is a form of vitamin E, and it can penetrate skin where it acts as a protective anti-oxidant.

*Surfactants*

**Sodium lauryl glucose carboxylate** is a gentle anionic surfactant that works best in mild acid conditions (pH around 5) and this explains why **malic acid** is also an ingredient.

**Lauryl glucoside** is a gentle but non-ionic surfactant.

*Antimicrobials*

**Sodium benzoate** is a powerful antibacterial agent.

**Methylisothiazolinone** (also known as MIT) is a more powerful germicide and kills all kinds of microbes.

There are other brands of intimate toilet wipes, such as Velvet® and Bella® as well as wipes for babies and toddlers' bottoms.

## 3.  HANDWASH

*The Chemistry*

Washing hands with soap and water is still an effective way to remove germs and dirt, even more effective if you wash, rinse, and then wash a second time. However, soap can be rather too good at removing natural oils from the skin and there are much gentler, but equally effective, products for washing your hands, and they can be formulated to include skin conditioners. They have the benefit of being dispensed in a more convenient way as a liquid, whereas soap tends to sit in a dish that becomes rather slimy.

Here we look at three kinds of handwash: a regular one that is labelled as antibacterial; a version that includes silver; and one that stresses it is nice to use and suggests it may even de-stress the user.

### Carex® Moisture Plus

*Contents*

Aqua
Sodium laureth sulfate
Glycerin
Lauramidopropyl betaine
Cocamidopropyl PG-dimonium
  chloride phosphate
Laureth-4
Polyquaternium-39
Lactic acid
Hexylene glycol
Parfum
Sine adipe lac
Styrene acrylates copolymer
PEG-40 hydrogenated castor oil
Trideceth-9
Pentylene glycol
Tetrasodium EDTA
Sodium chloride
Sodium benzoate
Hexyl cinnamal
Limonene
Phenoxyethanol
Methylparaben
Ethylparaben
Butylparaben
Propylparaben
Isobutylparaben.

**Aqua** is water and a solvent.

*Surfactants and water softener*

**Sodium laureth sulfate** is an anionic surfactant, and **lauramidopropyl betaine** is a non-ionic surfactant. However, **cocamidopropyl PG-dimonium chloride phosphate** is a cationic surfactant and, by virtue of this, it will leave behind a protective layer on the skin. **Laureth-4** also acts as a non-ionic surfactant but it has a dual role and works as an emulsifier.

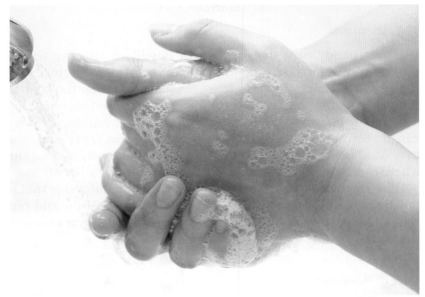

© Shutterstock

**Tetrasodium EDTA** counteracts hard water by sequestering calcium ions that would otherwise interfere with the action of the surfactants.

*Antibacterials*

The main antibacterial agents are **sodium benzoate** and **phenoxyethanol**. (The others are **methylparaben, ethylparaben, butylparaben, propylparaben**, and **isobutylparaben**, but these are best seen as preservatives that come with the various ingredients.)

*Moisturisers*

**Polyquaternium-39** is a positively charged polymer and it is by virtue of this charge that it adheres to the skin. **Glycerin** also keeps the skin moist.

*Emulsifiers*

There are several ingredients that act this way and they are **pentylene** and **hexylene glycol, styrene acrylates copolymer,**

**PEG-40 hydrogenated castor oil**, and **trideceth-9**. The last of these also spreads the Carex® more evenly over the skin.

**Lactic acid** keeps the pH slightly acidic.

**Sodium chloride** ensures that the contents do not lose water on standing and that the ingredients remain well mixed.

**Parfum**. Perfume raw material is a complex mixture, some items of which have to be identified in a label because some people may be sensitive to them. The ones so identified here are **hexyl cinnamal**, which smells of camomile and **limonene**, which smells of oranges.

**Sine adipe lac** (meaning 'without fat milk') is just a fancy way of saying skimmed milk.

## Carex® Antibacterial Handwash with Silver

*The Chemistry*

Silver is deadly to all microbes. As the silver ion, $Ag^+$, it attaches itself to the sulfur–sulfur bonds in their enzymes. This denatures these essential components of the cell and thereby kills them.

*Contents*

| | |
|---|---|
| Aqua | C11–15 pareth-40 |
| Sodium laureth sulfate | C11–15 pareth-7 |
| Cocamidopropyl betaine | Tetrasodium glutamate diacetate |
| Sodium chloride | Benzotriazolyl dodecyl *p*-cresol |
| Lactic acid | Propylene glycol |
| Glycerin | Limonene |
| Polyquaternium-7 | Linalool |
| Parfum | Hexyl cinnamal |
| Silver citrate | Citric acid |
| Sodium benzoate | CI 42051 |
| Styrene/acrylates copolymer | CI 60730 |

Several of these ingredients are explained above for Carex® Moisture Plus. Ones new to this formulation are as follows:

*Surfactants*

**Cocamidopropyl betaine** is a non-ionic surfactant.

**C11–15 pareth-40** and **C11–15 pareth-7** are non-ionic surfactants.

**Tetrasodium glutamate diacetate** counteracts hard water by sequestering calcium ions that would otherwise interfere with the action of the surfactants.

*Active Agents*

**Silver citrate** is the silver salt of citric acid and as such is soluble, especially in the presence of citric acid.

**Citric acid** ensures the handwash is slightly acidic, which is now seen as preferable for skin rather than alkaline products which remove more natural oils.

*Moisturisers*

**Polyquaternium-7** is attracted to skin and provides a protective film.

**Propylene glycol** is a skin conditioner.

**Parfum** is a mixture of fragrances, some natural, some artificial, and it may contain items that some individuals are sensitive to. These are identified as **limonene**, which smells of oranges; **linalool,** which has a floral spicy note; and **hexyl cinnamal**, which smells of camomile.

**Benzotriazolyl dodecyl *p*-cresol** is an antioxidant.

**CI 42051** is a blue dye and **CI 60730** is a violet dye.

## Palmolive® Ayuritual Joyous Liquid Handwash

This claims to offer 'body and mind harmony' on the basis that it contains plant extracts that are thought by some to have medical and spiritual benefits.

*Contents*

| | |
|---|---|
| Aqua | Citric acid |
| Sodium C12–13 pareth sulfate | Tetrasodium EDTA |
| Sodium laureth sulfate | BHT |
| Cocamidopropyl betaine | Benzophenone-4 |
| Sodium chloride | *Morinda citrifolia* fruit extract |
| Parfum | *Nelumbo nucifera* extract |
| Cocamide MEA | Butylphenyl methylpropional |
| Sodium salicylate | Citronellol |
| Sodium benzoate | Hexyl cinnamal |

**Aqua** is water and the solvent.

**Sodium C12-13 pareth sulfate** and **sodium laureth sulfate** are anionic surfactants.

**Cocamidopropyl betaine** and **cocamide MEA** are non-ionic surfactants with excellent foaming properties.

**Sodium chloride** is needed to keep the handwash stable as an emulsion.

**Parfum**. Perfume raw material is a complex mixture, some items of which have to be identified in a label because some people may be sensitive to them. The ones so identified here are **butylphenyl methylpropional,** which is an artificial fragrance that has a strong floral note; **citronellol,** which smells of geraniums and roses, and **hexyl cinnamal,** which smells of camomile.

**Sodium salicylate** acts as an anti-inflammatory agent.

**Sodium benzoate** keeps the handwash free of germs.

**Citric acid** ensures the handwash is slightly acidic, which is now seen as preferable for skin rather than alkaline products which remove more natural oils.

**Tetrasodium EDTA** counteracts hard water by sequestering calcium ions that would otherwise interfere with the action of the surfactants.

**BHT** (butylated hydroxytoluene) is a powerful antioxidant.

**Benzophenone-4** protects against UV rays.

*Morinda citrifolia* (also known as Indian mulberry) **fruit extract** is believed to act as an anti-inflammatory on the skin.

*Nelumbo nucifera* (also known as Indian lotus) **extract** is an alternative medical treatment for dermatitis.

Other hand wash products are available, such as those under the brand names of Dove® and Dettol®, as well as several up-market and own-label brands.

## 4. HAND CREAMS

*The Chemistry*

Natural oils are lost from the skin on washing and may need replacing or supplementing during cold or windy weather, when the
skin becomes very dry, chapped, and even painfully cracked. These
natural oils can be replaced by oils derived from animals and
plants. Such products need to include emulsifiers and emollients,
and remain germ-free by including antimicrobial agents.

### Nivea® Creme

*Contents*

| | |
|---|---|
| Aqua (water) | Magnesium sulfate |
| Paraffinum liquidum | Magnesium stearate |
| Cera microcristallina | Parfum (fragrance) |
| Glycerin | Limonene |
| Lanolin alcohol (Eucerit®) | Geraniol |
| Paraffin | Hydroxycitronellal |
| Panthenol | Linalool |
| Decyl oleate | Citronellol |
| Octyldodecanol | Benzyl benzoate |
| Aluminium stearates | Cinnamyl alcohol. |
| Citric acid | |

*Oils and Waxes*

**Paraffinum liquidum** is also known simply as liquid paraffin or
mineral oil. It is a colourless mixture of long chain hydrocarbons and comes from oil. It acts partly as a solvent for the
other ingredients and remains as a waterproof layer, which
reduces moisture loss from the skin, as does **Cera microcristallina,** which is a wax of branched hydrocarbon chains
and is extracted from oil. Its crystals are much smaller than
those of normal hydrocarbon waxes.

*Emollients*

**Lanolin alcohol (Eucerit®)** is a complex combination of organic alcohols obtained by the hydrolysis of lanolin.
**Decyl oleate** makes the skin feel smooth. It is a derivative
of oleic acid, which comes mainly from olive oil, combined
with the alcohol 1-decanol, which comes mainly from
coconut oil.

**Octyldodecanol** acts as an emollient. It is a long chain fatty alcohol.

**Glycerin** is a humectant that prevents moisture loss.

**Paraffin** consists of liquid hydrocarbons.

**Panthenol** is a precursor of vitamin B5 and the body can convert it to this vitamin. It is thought to have beneficial effects for the skin.

**Aluminium stearates** prevent the ointment from becoming hard.

**Citric acid** ensures the hand cream is slightly acidic, which is now seen as preferable for skin rather than alkaline products, which react with natural oils.

**Magnesium sulfate** and **magnesium stearate** provide magnesium, which is thought to be good for the skin.

**Parfum (fragrance).** Perfume raw material is a complex mixture some items of which have to be identified in a label because some people may be sensitive to them. The ones so identified here are **limonene**, which smells of oranges; **geraniol**, which smells of roses; **hydroxycitronellal**, which smells of lime; **linalool**, which has a floral/spicy note; **citronellol**, which smells of geraniums and roses; and **cinnamyl alcohol**, which has a floral/spicy smell.

**Benzyl benzoate** acts as an antimicrobial agent.

## Neutrogena® Norwegian Formula® (Unscented)

This brand prides itself on being able to cope with the most severe conditions of weather, as might be expected in a Norwegian winter.

*Contents*

| | |
|---|---|
| Aqua | Sodium cetearyl sulfate |
| Glycerin | Dilauryl thiodipropionate |
| Cetearyl alcohol | Sodium sulfate |
| Stearic acid | Methylparaben |
| Palmitic acid | Propylparaben |

**Aqua** is water and acts as a solvent.

**Glycerin** is a humectant that bonds to the skin and prevents moisture loss, and will even attract moisture from the atmosphere.

**Cetearyl alcohol** acts as a thickener and is an emollient so keeps the ingredients well mixed.

**Stearic acid** is a long chain saturated fatty acid that comes mainly from animal fats and acts as an emulsifier.

**Palmitic acid** is also a long chain fatty acid and this comes mainly from palm oils. It too acts as an emulsifier.

**Sodium cetearyl sulfate** is a mild anionic surfactant that also acts as an emulsifier.

**Dilauryl thiodipropionate** is an antioxidant designed to preserve the ingredients and protect the skin.

**Sodium sulfate** helps keep the viscosity of the contents stable.

**Methylparaben** and **propylparaben** are antibacterial agents.

There are many other hand creams, such as the brands Dove®, Vaseline®, Boots, Crabtree & Evelyn®, which are the kind you keep by the sink. There are also skin creams that are really cosmetics, and these are described in Chapter 7.

## 5. ABSORBENT PRODUCTS

*The Chemistry*

These products rely on a superabsorbent polymer (SAP), which is poly(acrylic acid). They are used by babies, women, the incontinent, and astronauts. The average adult passes between 800 and 2000 ml of urine a day depending on their food and fluid intake. SAP can absorb vast amounts of water, swelling as it does so. Indeed, the 10 grams of polymer in a typical nappy can hold up to 50 times its weight of water.

Some of the groups attached to the SAP are salt groups, which are ionic, and this greatly increases the ability of the polymer to absorb water. In the commercial product, the polymer also includes a small amount of a cross-linking agent, which is necessary to make the polymer insoluble. (The more cross-links there are, the more insoluble the polymer becomes, but then it becomes less able to swell and so less able to absorb water.)

## Pampers® and Tena®

During the course of a day, a baby will pass around 500 ml of urine. In theory, a single disposable nappy could absorb all of it. It has four components:

(1) The body-side liner, which is made from polypropylene fabric because this feels soft against the skin. This layer also has two flaps running along its length, which act as a barrier to prevent urine flowing outwards.
(2) The surge layer, which is designed to disperse the urine from its point of impact. This layer is usually made of cellulose.
(3) The all-important absorbent layer, which consists of a mixture of the SAP dispersed in cellulose fluff.
(4) The outer covering, again made of waterproof polypropylene with a cloth-like feel. This layer also has built into it the stretchy ears with which to fasten the nappy and elastic at the outsides to grip the legs, thereby preventing excess urine from leaking. This outer layer may also be made permeable, with millions of tiny holes allowing air to circulate.

Adults too can suffer from occasional incontinence and there are products to cope with this condition, such as TENA® and there are versions for both women and men. They also rely on a SAP.

### Always® Panty Liners

These consist of a plastic film layer in contact with the body. This is made of polypropylene, a polymer that feels soft against the skin, and it has been made permeable. Below the outer layer is the SAP, which absorbs any discharge and retains it. The SAP is mixed with cellulose so that it is evenly spread throughout the pad.

Finally, there is the outer layer, which has to be waterproof, and this is a layer of polythene.

Another popular brand of panty liners is TENA® and these rely on the same SAP.

## 6.  SUPPOSITORIES

*The Chemistry*

You may need to use a suppository to obtain relief when constipated, and this may simply be a bullet-shaped gel of glycerol jelly which is moistened and inserted into the anus. The warmth of the body melts the gel and the glycerol it contains then acts as a lubricant to ease evacuation of the bowels. These kinds of suppository will be effective within 15–30  minutes.

Another reason to use a suppository is if you suffer from haemorrhoids (also known as piles), which are enlarged blood vessels. This is the more serious condition because the swollen blood vessels may bleed when you pass a motion and may even be so large they protrude from the anus. Piles can be treated with the following kind of suppository.

### Anusol®

*Contents*

| | |
|---|---|
| Zinc oxide | Hard fat |
| Bismuth subgallate | Kaolin light |
| Balsam Peru | Titanium dioxide |
| Bismuth oxide | Miglyol 812 |

**Zinc oxide,** ZnO, is used to treat all kinds of skin conditions. It acts as an astringent – in other words it tends to shrink the skin a little – and in this case it reduces the size of the swollen blood vessels. Zinc also has some antibacterial activity.

**Bismuth subgallate** and **bismuth oxide** act to protect the passage leading to the anus by suppressing bacteria and these will also to shrink the piles a little.

**Balsam Peru** is a traditional antiseptic and is reputed to reduce inflammation, and it helps healing. It also relaxes the muscles of the anus making passing faeces easier.

**Hard fat** is lard and it bulks out the suppository and acts as a lubricant.

**Kaolin light** is the mineral known as china clay (aluminium silicate), finely ground. It may act to reduce swelling in the large intestine.

**Titanium dioxide** is a white pigment.

**Miglyol 812** is caprylic/capric triglyceride and it acts to soften the skin.

There are also Germoloids®, pharmacist own-brand suppositories, and glycerine-based ones such as Care + ®.

## 7. PREGNANCY TEST

*The Chemistry*

Pregnancy tests rely on detecting a hormone called human chorionic gonadotropin (hCG), which is released when a woman is pregnant and is excreted in her urine. It is only produced after the fertilised egg has become implanted in the uterus and that occurs about a week after fertilisation. There are several pregnancy test kits and most are 99% reliable.

### First Response® Pregnancy Test Kit

If the user sees two red lines, she is pregnant; if there is only one, then she is not.

*Components*

| | |
|---|---|
| Airtight foil wrapper | Immunoassay strip |
| Desiccant | Antibodies |
| Plastic housing | Gold nanoparticles |
| Absorbent wick | |

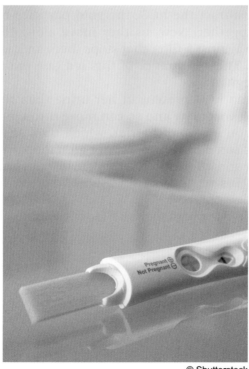

© Shutterstock

The **airtight foil wrapper** keeps the device dry before use. If the wrapper is removed then the test kit must be used within 12 hours.

**Desiccant** is silica gel and is designed to ensure that the device remains perfectly dry.

The **plastic housing** is needed to protect the test strip and it consists of a coloured polymer with an inbuilt polycarbonate window where the test results are shown.

**Absorbent wick** is the target for the stream of urine, which must impinge on it for at least 10 seconds. It absorbs the urine and this is then conveyed by capillary action into the body of the device where it comes in contact with the analysis strip.

**Immunoassay strip** consists of compressed polymer fibres made of rayon or polyester. The urine travels along it and in so doing it encounters **antibodies**. These are small fragments of protein designed to identify hCG and are attached to **gold nanoparticles**. These appear red because of the way nanoparticles of gold interact with light.

When they attract the hCG, the gold particles become larger and cluster together. Now, as they move along the fibre strip, they come up against a filter that stops them and, as it does so, it becomes red. The appearance of this red line indicates that the woman is pregnant.

If there is no hCG present in her urine then the gold nanoparticles don't coagulate and so can move through the first filter to be caught on a second barrier, which then shows red. Even when there is some hCG present, not all nanoparticles will have picked it up and those that haven't will pass through to the second site. This second red line shows the test has been successfully carried out.

Another popular pregnancy test kit is Clearblue®.

## 8.  AIR FRESHENERS

*The Chemistry*

There are five categories of domestic smell that humans find most offensive: toilet odours, tobacco smoke, stale cooking smells, pets, and mould.

Toilet odours, and especially those due to other people, touch a primeval nerve that registers disgust. It is part of our primate origins that while we may not find the smell of our own body's emissions unpleasant, others will find them objectionable. Toilet odours are made up of a few strong smelling molecules such as butanoic and 3-methylbutanoic acids, which are formed by bacteria in the gut, plus some $H_2S$, along with the nitrogen-containing chemical skatole, which is particularly characteristic of human faeces.

One way to dispel an offensive odour is to mask it with nicer smelling odours so that together they form a mixture of

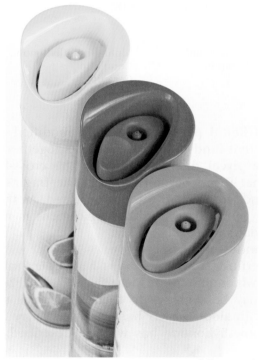

© Shutterstock

fragrance notes that smells pleasant, rather like the way that perfumes contain both top and base notes – see Chapter 7, The Bedroom. AirWick®, the one described below, acts this way.

Another way is to remove the offensive molecules from the air and this can be achieved by a spray containing triethylene glycol, which is particularly good at absorbing molecules as its droplets of vapour move through the air and eventually fall to the floor. Oust® works like this.

The third way is to encircle the molecules within cyclodextrin molecules, so your nose can no longer detect them, and this is the basis of Febreze®.

### Air Wick Crystal Air® Yorkshire Dales, White Roses and Pink Sweet Pea Aroma

This air freshener is supposed to mimic the smells associated with an outdoor location and the flowers described. It works by slowly releasing the fragrance molecules which the diffuse into the air so as to mask the offending odours.

*Contents*

| | |
|---|---|
| Maleinised liquid polybutadiene | Alpha-isomethylionone |
| Triethylene glycol diamine | Coumarin |
| Benzophenone-3 | Geraniol |
| Parfum | Citronellol |
| Linalool | Limonene |

**Maleinised liquid polybutadiene** is a modified soft rubbery material that binds all the ingredients together.

**Triethylene glycol diamine** provides the cross-linking molecules that form the gel in which the fragrances are dissolved and from which they are released. As the scents are lost, the gel shrinks until a replacement is needed. This takes about six weeks.

**Benzophenone-3** is a preservative.

**Parfum.** Perfume raw material is a complex mixture, some items of which have to be identified in a label because some people may be sensitive to them. The ones so identified here are **linalool**, which has a flowery/spicy smell, while **alpha-isomethylionone** has what is described as a 'clean' smell, **coumarin** has the smell of new-mown hay, **geraniol** smells of roses, **citronellol** smells of geraniums and roses, and **limonene** smells of oranges.

## Oust®

This air freshener works by spraying a mist that absorbs malodorous molecules and carries them down to floor level.

*Contents*

n-Alkyl dimethyl benzyl ammonium saccharinate
Triethylene glycol
Propane
Isobutane
Ethanol
Water

**n-Alkyl dimethyl benzyl ammonium saccharinate** is a cationic surfactant that also has disinfectant properties. Its positive charge is counterbalanced by the negative form of saccharin and this helps solubilise it. Its role is to ensure that all the contents remain blended and don't separate out in the can.

**Triethylene glycol** is particularly good at absorbing molecules and removing them from the air, and this product contains around 5%. It is especially good at absorbing those molecules that have a sulfur atom or an aromatic ring as part of their structure, and these are part of lavatory-type smells.

**Propane** and **isobutane** are gases that act as propellants. These are under pressure in the canister and they force out the content through a nozzle when the pressure is released.

**Ethanol** and **water** are solvents.

## Febreze® Air Effects

This contains molecules that trap malodourous ones like $H_2S$ and skatole so they cannot be detected by the nose.

*Contents*

Nitrogen
Purified water
Alcohol
Hydrogenated castor oil
Dialkyl sodium sulfosuccinate
Polyacrylate
Hydrochloric acid
Benzisothiazolinone
Various perfume blends
Cyclodextrin

**Nitrogen** is the propellant gas. It is under pressure in the canister and forces out the content through a nozzle when the pressure is released.

**Purified water** and **alcohol** are the solvents for the various ingredients.

**Hydrogenated castor oil** and **dialkyl sodium sulfosuccinate** act to keep all the ingredients nicely blended as an emulsion.

**Polyacrylate** attracts water, which ensures that the cyclodextrin with its trapped molecules does not dry out and release them again.

**Hydrochloric acid** keeps the liquid slightly acidic and it will also combine chemically with some amine-based offensive odour such as skatole and deactivate them.

**Benzisothiazolinone** is a preservative.

**Various perfume blends** are unspecified but the product is called 'cotton fresh' and it leaves a lingering smell associated with laundered linen.

**Cyclodextrin** is the active agent that captures foul-smelling molecules.

# CHAPTER 6

# The Cupboard Under the Stairs

© Shutterstock

Chemistry at Home: Exploring the Ingredients in Everyday Products
By John Emsley
© John Emsley, 2015
Published by the Royal Society of Chemistry, www.rsc.org

This is where we keep those products that we use occasionally.

1. Lubrication (WD40®)
2. Shoe cleaning (Kiwi® Shoe Polish, Kiwi® Quick 'n' Clean Wipes)
3. Metal cleaning (Silvo® Tarnish Guard, Silvo® Wadding, Goddard's® Long Term Silver Polish Cloth, Brasso® Metal Polish, Brasso® Wadding)
4. Polish (Mr Sheen® Multi-Surface Polish Aerosol)
5. Glue (Loctite® GO2®)
6. Wallpaper paste (Solvite®)
7. Pet products (Bob Martin® FleaClear®)

## 1. LUBRICATION

*The Chemistry*

Things sometimes stick when we want them to move easily. Screws, nuts and bolts, hinges, door handles, locks, gears, shears, and axles can all cause problems. Then we need some lubrication, and the best kind of chemical is a hydrocarbon grease or oil that can reduce the friction between the two surfaces.

If the gap between the two surfaces is not easily accessible then a penetrating oil is needed in which the lubricants are dissolved in a low viscosity solvent that can then carry them to where they are needed.

## WD40®

This fluid is a combination of hydrocarbon-based solvents and various oils, including heavy duty ones. Although the makers of WD40® don't reveal its contents, these can be gleaned from various material safety information sites in countries where the key information has to be revealed by law, such as Canada. WD40® does not contain either Teflon® type chemicals or silicone oils, which were once suspected of being its secret ingredients.

*Contents*

| | |
|---|---|
| Paraffin | Transformer oil |
| White spirit | Vacuum pump oil |
| Mineral oil | Dimethyl naphthalene |
| Lubricating oil | Carbon dioxide |

The hydrocarbons are listed in order of their viscosity, the more fluid ones acting as solvents for the ones that do the real lubricating. All these are produced by distilling crude oil.

**Paraffin** is a mixture of short chain and volatile hydrocarbons.

**White spirit** is mixture of less volatile hydrocarbons and it accounts for two thirds of WD40®.

**Mineral oil** (also known as liquid paraffin) consists of non-volatile hydrocarbons.

**Lubricating oil** comprises longer hydrocarbon chains with side chains attached. It is produced when crude oil is distilled between 300 and 370 °C.

**Transformer oil** is used as an insulating fluid in high voltage electrical transformers.

**Vacuum pump oil** has to have a low vapour pressure so only the least volatile hydrocarbon liquids are used.

**Dimethyl naphthalene** is a hydrocarbon solvent that boils at around 260 °C. It has two co-joined benzene rings and this explains why WD40® burns with a smoky flame.

**Carbon dioxide** is the gas that acts as the propellant. It is under pressure in the canister and forces out the content through a nozzle when the pressure is released.

Another brand of penetrating lubricating oil is 3-in-One®.

## 2. SHOE CLEANING

*The Chemistry*

Leather shoes often need cleaning, and the polish designed for this purpose has to keep the leather supple, make it waterproof, and maybe revive its shine. Thus it needs to contain a mixture of oils and waxes in a hydrocarbon solvent base. The traditional tin of shoe polish is still the kind that is most bought, although shoe creams and aerosol kinds are now as popular.

### Kiwi® Shoe Polish (Black)

*Contents*

| | |
|---|---|
| Lanolin | Naptha |
| Turpentine | Ethylene glycol |
| Carnauba wax | Carbon black or nigrosine |
| Gum Arabic | |

**Lanolin** comes from wool. It is both waterproofing and keeps the leather supple.

**Turpentine** is a natural hydrocarbon solvent.

**Carnauba wax** consists of long chain fatty acid esters and alcohols. It is used in polishes for things other than shoes, such as cars, furniture, and lipstick.

**Gum Arabic** comes from the acacia tree and gives the polish the right consistency.

**Naptha** is a mixture of hydrocarbons and a solvent; it evaporates as the shoe is brushed.

**Ethylene glycol** prevents the shoe polish from drying out.

**Carbon black** is essentially soot; **nigrosine** is an aniline-based black dye.

Other brands of shoe polish are Cherry Blossom®, G wax®, and Loake®.

## Kiwi® Quick 'n' Clean Wipes

If you haven't time to spend cleaning shoes, then a quick wipe with this product will achieve the same result – almost.

*Contents*

| | |
|---|---|
| Aqua | Sodium benzoate |
| Polyester and viscose | Parfum |
| Propylene glycol | Citric acid |
| Phenoxyethanol | *Aloe Barbadensis* leaf juice |
| PEG-40 hydrogenated castor oil | |

**Aqua** is water and acts as a solvent.

**Polyester and viscose** are the fibres from which the cloth is made.

**Propylene glycol** is a solvent like alcohol but not as volatile.

**Phenoxyethanol** and **sodium benzoate** are antibacterial agents.

**PEG-40 hydrogenated castor oil** acts as a surfactant and emulsifier.

**Parfum** is not specified, so contains no molecules to which some people might be sensitive.

**Citric acid** ensures the wipes are slightly acidic and this helps protect the hands.

*Aloe Barbadensis* **leaf juice** is the liquid more commonly known as aloe vera and is added to protect the hands while wiping the shoes.

### 3. METAL CLEANING

*The Chemistry*

Copper, brass, chrome, and silver can become dull due to the formation of an oxide and sulfide layer on their surface. This layer can be removed by rubbing with a gentle abrasive in a suitable liquid base. In some cases the liquid is absorbed on wadding for convenience.

Silver is the most prone to darkening and the following products offer protection against future discolouration. They leave behind a chemical that actually incorporates a sulfur atom to which is attached a long chain hydrocarbon. The sulfur is attracted to the surface of the metal but does not darken it, while the hydrocarbon layer acts as an invisible barrier against other possible airborne molecules that can attack it.

Some metal polishes are designed to clean silver, some to clean brass.

## Silvo® Tarnish Guard

*Contents*

Aqua
Quartz/kaolinite mixture
Isopropyl alcohol
Octadecane-1-thiol
Calcium carbonate

Parfum
Polysorbate 60
Benzyl salicylate
Limonene

© Shutterstock

**Aqua** is water and acts as a solvent.

**Quartz/kaolinite mixture** is a mild abrasive designed to dislodge the discolouring layer and smooth out any scratch marks.

**Isopropyl alcohol** is a solvent for some of the ingredients.

**Octadecane-1-thiol** this sulfur-containing molecule acts to protect the cleaned surface in the manner described above.

**Calcium carbonate** makes an excellent scouring powder and yet is one that does not scratch metal surfaces.

**Parfum.** Perfume raw material is a complex mixture, some items of which have to be identified in a label because some people may be sensitive to them. The one so identified here is **limonene**, which has a slight orangey smell.

**Polysorbate 60** is a surfactant and an emulsifier.

**Benzyl salicylate** acts as a fixative.


### Silvo® Wadding

*Contents*

C8–10 alkane/cycloalkane/aromatic hydrocarbons
Quartz
Tallow acid

Octadecane-1-thiol
Ammonium hydroxide
Aqua
Colorant

**C8–10 alkane/cycloalkane/aromatic hydrocarbons** is a blend of several kinds of hydrocarbon solvent: short chain ones, cyclic ones, and some that incorporate a benzene ring, which is what the term 'aromatic' means.

**Quartz** is silicon dioxide ground very finely and is a gentle abrasive.

**Tallow acid** is a mixture of fatty acids derived from tallow, which is hard animal fat from cows and pigs. It acts as an emulsifier.

**Octadecane-1-thiol** this sulfur-containing molecule acts to protect the cleaned surface in the manner described above.

**Ammonium hydroxide** acts as a buffering agent to keep the wadding slightly alkaline. (Metals are prone to attack by acids.)

**Aqua** is water.

**Colorant** is not specified.

## Goddard's® Long Term Silver Polish Cloth

This product carries a royal warrant and is similar to the above in that it relies on octadecane-1-thiol to protect the cleaned silver.

The other ingredients are **aliphatic hydrocarbons,** which act as the solvent, **perfume,** which is not specified and so contains nothing to which people might be sensitive, and both **sodium *p*-choro-*m*-cresol** and **sodium *o*-phenylphenate,** which act as antibacterial agents.

## Brasso® Metal Polish

*Contents*

| | |
|---|---|
| Kerosine (petroleum) hydro-<br>  desulfurized | Fatty acids, C14–18 and C16–18<br>  unsaturated |
| Silicon dioxide | Aqua |
| Kaolinite | Ammonium hydroxide |
| Kaolin | |

**Kerosine (petroleum) hydro-desulfurized** is paraffin that has been freed of sulfur compounds.

**Silicon dioxide** is finely ground quartz, **kaolinite** is the mineral aluminium silicate and **kaolin** is china clay, whose main component is kaolinite. These act as gentle abrasives.

**Fatty acids, C14–18 and C16–18 unsaturated** act as emulsifiers.

**Aqua** is water.

**Ammonium hydroxide** ensures that the polish does not become acidic. (Metals are prone to attack by acids.)

## Brasso® Wadding

*Contents*

| | |
|---|---|
| C8–10 alkane/cycloalkane/aromatic<br>  hydrocarbons | Kaolinite |
| Quartz | Ammonium hydroxide, |
| Fatty acids, C14–18 and C16–18<br>  unsaturated | Iron hydroxide |

**C8-10 alkane/cycloalkane/aromatic hydrocarbons** is a blend of several kinds of hydrocarbon solvent: short chain ones, cyclic ones, and some that incorporate a benzene ring, which is what the term 'aromatic' means.

**Quartz** is silicon dioxide ground very finely to act as a gentle abrasive, along with **kaolinite**, which is the mineral aluminium silicate.

**Fatty acids, C14–18 and C16–18 unsaturated** act as emulsifiers.

**Ammonium hydroxide** ensures that the polish does not become acidic. (Metals are prone to attack by acids.)

**Iron hydroxide** is probably hydrated iron oxide, which is known as jeweller's rouge and used to polish soft metals and glass. It also protects the fatty acids against oxidation.

## 4.  POLISH

*The Chemistry*

At times there is a need to clean and polish household surfaces that have become dull, such as furniture and woodwork. Then we need a fluid that will lift the dirt onto a cloth and leave behind a clean, shiny surface, plus a layer of something to protect the surface.

## Mr Sheen® Multi-Surface Polish Aerosol

*Contents*

| | |
|---|---|
| C10–12 alkane/cycloalkane | Hexyl cinnamal |
| Butane | Butylphenyl methylpropional |
| Isobutane | Benzyl salicylate |
| Propane | Microcrystalline wax |
| Dimethicone | Paraffin |
| Sorbitan oleate | Sodium benzoate |
| Parfum | |

*Solvents*

> **C10–12 alkane/cycloalkane** is a mixture of linear and cyclic liquid hydrocarbons.
> **Paraffin** is a mixed hydrocarbon with slightly longer carbon chains.

*Polishes*

> **Dimethicone** is a silicone oil that leaves behind a protective, water-repelling layer on the surface.
> **Microcrystalline wax** consists of tiny crystals of long chain hydrocarbons. It provides a shiny surface.
> **Butane, isobutane**, and **propane** are propellants. These are under pressure in the canister and they force out the content through a nozzle when the pressure is released.
> **Sorbitan oleate** is a mild surfactant. It is also designed to ensure all the ingredients form a homogeneous mixture.
> **Parfum**. Perfume raw material is a complex mixture, some items of which have to be identified in a label because some

people may be sensitive to them. The ones so identified here are **hexyl cinnamal,** which smells of camomile; and **butyl-phenyl methylpropional**, which has a floral smell.
**Benzyl salicylate** is a fixative. Of itself, it has almost no odour.
**Sodium benzoate** is a powerful germicide.

Another popular brand of polish is Pledge®.

## 5.   GLUE

*The Chemistry*

We may be reluctant to throw away something that gets broken, like a treasured ornament or the handle of a favourite mug. Then we need a strong glue.

A glue is designed to adhere to the two surfaces to be joined and then to ensure these remain attached. This can be achieved if they are linked by polymer chains. Previous generations relied on the natural polymer keratin. This is the polymer of hair, nails, and animal hooves and horns, and it was the last of these that provided the commercial product. Today we generally use the polymers produced by the chemical industry.

## Loctite® GO2®

This will stick together all combinations of wood, metal, leather, and glass, and certain plastics but must not be used for polythene, Teflon®, or polystyrene because these may be weakened by the benzene-based solvent in this glue or, in the case of Teflon, even this glue will not form a permanent bond to its surface. Loctite® GO2® begins to set once it is in contact with traces of moisture, such as that which clings to all surfaces or comes from the atmosphere. Spread a little glue on the two surfaces to be joined, press them together, and leave for 30 minutes for the glue to set.

*Contents*

Polyurethane
Benzene C10–13 alkyl derivatives
3-(Trimethoxysilyl)propylamine

Trimethoxyvinylsilane
Methanol

**Polyurethane** is the polymer that is going to form the bond linking the surfaces to be joined.

**Benzene C10–13 alkyl derivatives** are hydrocarbon solvents that together make up around 30% of the contents of the adhesive. These are liquids in which a hydrocarbon chain is attached to a benzene ring and with boiling points in excess of 250 °C.

**3-(Trimethoxysilyl)propylamine** accounts for around 5% of the product. It helps the glue bond to surfaces and is especially good at sticking plastics to glass or metals.

**Trimethoxyvinylsilane** accounts for around 5% of the product. It forms cross-links between the polyurethane polymer chains and is known as an adhesion promoter.

**Methanol** is a volatile solvent.

Other popular ranges of adhesives go under the brand names of Evo-Stik® and Araldite®.

## 6. WALLPAPER PASTE

*The Chemistry*

Wallpaper paste either comes as a starch-based powder, which is how most DIY wallpapering is done, or as a ready-mixed cellulose paste, which is what professionals use. Both are carbohydrate polymers, and while cellulose is not a breeding ground for moulds, starch can be, so it has to be chemically modified to make it unsuitable for moulds to grow.

### Solvite®

This is a modified-starch wallpaper paste.

*Contents*

Potato starch carboxymethyl ether sodium salt
Sodium hydroxide

*N*-Butyl-1,2-benzisothiazolin-3-one
Polymeric biguanide hydrochloride

**Potato starch carboxymethyl ether sodium salt** is starch that has been reacted chemically with sodium hydroxide and monochloroacetic acid to make it impervious to breakdown

© Shutterstock

by chemical or microbial attack so it will give a long-lasting bond between wallpaper and wall.

**Sodium hydroxide** accounts for less than 0.1% of the product, but it is enough to make the paste slightly alkaline.

*N*-**Butyl-1,2-benzisothiazolin-3-one**, of which there is less than 0.5%, acts as the fungicide to prevent moulds forming on the paste.

**Polymeric biguanide hydrochloride**, of which there is less than 0.1%, also acts as a fungicide as well as protecting the paste against insects.

There are other brands of wallpaper paste, such as Mangers®.

## 7.  PET PRODUCTS

*The Chemistry*

Household pets are prone to infestations by fleas (*Ctenocephalides canis*) and ticks (*Rhipicephalus sanguineus* and *Ixodes recinus*). They need to be killed and there are products for doing this. These must contain an insecticide, and it has to be one that will do the job but without being a threat to the health of the pet or to humans. Chemists have found a range of such insecticides.

### Bob Martin® FleaClear®

This will rid your dog of fleas and ticks within 24 hours. The product comes as a capsule. You break the end off and then squeeze the liquid on to the skin at the back of the dog's neck, so there is no risk of the animal licking it.

*Contents*

| | |
|---|---|
| Fipronil | Benzyl alcohol |
| Butylhydroxyanisole E320 | Diethylene glycol |
| Butylhydroxytoluene E321 |   monoethyl ether |

**Fipronil** is the insecticide that kills the fleas and there are only 0.134 grams (134 mg) in each capsule. (This insecticide is also used against pest infestations on golf courses and other such turf.)

**Butylhydroxyanisole** and **butylhydroxytoluene** are antioxidants, and these are added to protect the fipronil.

**Benzyl alcohol** acts partly as a solvent for the insecticide but also prevents the growth of bacteria in the product.

**Diethylene glycol monoethyl ether** is the solvent.

# CHAPTER 7

# The Bedroom

© Shutterstock

The products we use in the bedroom involve chemistry in a very personal way. They determine the way we look and the way we make love.

How we look is how the world sees us, and after some more image-conscious people have spent time in their bathrooms they may spend as much time at the dressing table using the products that cosmetic and fragrance chemists have devised. There are thousands of these, designed to make people look younger and more attractive. In this chapter we will look at typical ones such as foundation, lipstick, mascara and nail varnish. Then we will

Chemistry at Home: Exploring the Ingredients in Everyday Products
By John Emsley
© John Emsley, 2015
Published by the Royal Society of Chemistry, www.rsc.org

look at a face cream that offers additional benefits, and a luxurious body cream.

There are times in this room when the other kind of chemistry, sex, comes to the fore and again chemistry may be there to enhance our enjoyment of the great game of life.

The kinds of products in this chapter are as follows:

1. Foundation (Max Factor® Lasting Performance Foundation)
2. Lipstick (Lancôme® L'Absolu Rouge® Lipstick)
3. Mascara (L'Oréal® Volume Million Lashes® Extra-Black; Rituals® Perfection)
4. Nail polish (L'Oréal® Paris Colour Riche Nails Femme Fatale)
5. Anti-wrinkle cream (Boots No.7 Protect and Perfect® Intense Beauty Serum)
6. Perfume (Insolence® Eau de Parfum by Guerlain® of Paris)
7. Body cream (Rituals® Touch of Happiness)
8. Hair products (Silvikin®, L'Oréal® Elnett® Hairspray, Brylcreem®, Regaine® Foam Extra Strength)
9. Sex (Levonelle®, Condoms, Viagra®, K-Y® Jelly)

## 1.  FOUNDATION

*The Chemistry*

As we get older, our hair and skin change, and indeed these changes start to appear earlier in life than we would like. Most of us want to appear younger than we really are and there are many products that can be used for this. In Chapter 3 we saw how chemists could help disguise changes to our hair. In this section we will consider products for the skin, and again chemistry comes to our aid.

We need to protect our skin and enhance it and this is the basis of foundation make-up. The most important ingredients are silicones and masking agents. Silicones provide a shield against the weather and atmospheric pollution, and reduce moisture loss from the skin.

Masking agents will employ pigments to hide imperfections, while emollients will make the skin appear smoother and younger. We also need to protect the skin against the environment and to preserve these products for as long as possible. And they will need to have the right consistency and that means using viscosity improvers. Getting the right balance for all these factors is what foundation is all about.

© Shutterstock

## Max Factor® Lasting Performance Foundation

*Contents*

Aqua
Cyclomethicone
Titanium dioxide
Propylene glycol
Dimethicone
Talc
Aluminium starch octenylsuccinate
Dimethicone copolyol
Sodium chloride
PVP
Laureth-7
Arachidyl behenate
Sodium dehydroacetate
Trihydroxystearin
Propyl paraben

Methicone
Methyl paraben
Synthetic wax
Silica
Cetyl dimethicone copolyol
Polyglyceryl-4 isostearate
Hexyl laurate
Ethylene/methacrylate copolymer
Ethylene brassylate
Sodium acetate
Aluminium hydroxide
Stearic acid
Isopropyl titanium triisostearate
CI 77491, CI 77492, CI 77499

**Aqua** is water and acts as a solvent.

*Silicones*

There are several of these in this product. **Cyclomethicone** is a light silicone oil noted for its fluidity so it flows smoothly over the skin. **Dimethicone** and **methicone** are the more common types of silicone oil. These are more viscous and offer a more permanent protective layer. **Dimethicone copolyol** and **cetyl dimethicone copolyol** are silicones with other polymer chains attached, and these cling to the skin and protect it against moisture loss.

*Pigments*

**Titanium dioxide** is a white pigment with excellent covering power. It also protects against UV.

**Talc** is a natural white magnesium silicate mineral renowned for its softness and has been the basic ingredient of make-up for many years.

The product also contains a mixture of three iron oxides: **CI 77491**, which is brown; **CI 77492**, which is yellow; and **CI 77499**, which is black. Together they create a flesh-like tone.

**Isopropyl titanium triisostearate** is used as a support for the pigments in the foundation.

*Emollients*

**Propylene glycol** acts as a solvent for some of the ingredients and prevents the product from drying out. It also acts a moisturiser.

**Arachidyl behenate, trihydroxystearin,** and **hexyl laurate** are emollients.

*Viscosity Control*

**Aluminium starch octenylsuccinate** acts to moderate the viscosity.

**Sodium chloride** is common salt, which thickens the moisturiser to the right consistency. **Laureth-7** acts as a thickener.

**Synthetic wax** is the wax extracted from oil.

*Emulsifier*

**PVP** is short for polyvinylpyrrolidone and is a water-soluble polymer that holds the other ingredients together with a fluid consistency.

**Polyglyceryl-4 isostearate** is an emulsifier and keeps the ingredients smoothly blended.

*Preservatives*

**Sodium dehydroacetate** is a preservative.

**Propyl paraben** and **methyl paraben** are antibacterial agents.

*Fillers*

**Silica** and **aluminium hydroxide** are fillers.

*Protection*

**Ethylene/methacrylate copolymer** forms a transparent protective film on the surface of the skin.

*Fragrances*

**Ethylene brassylate** is a synthetic musk fragrance.

*Buffering Agent*

**Sodium acetate** and **stearic acid** work together to keep the pH of the foundation stable.

Other foundations are available under brand names such as L'Oreal® and Rimmel® and, like Max Factor®'s Lasting Performance Foundation, they offer several different shades to suit all skin types.

## 2. LIPSTICK

*The Chemistry*

Women have rouged their lips since the time of ancient Egypt, and there has always been a need to protect the lips whose layer of skin is thin and vulnerable to the weather.

Applying such colouring and protection took a big step forward with the introduction of push-up lipsticks in metal containers, which appeared in 1915 in the USA and were invented by Maurice Levy. Even so, the earliest lipsticks had many faults: they were greasy and offered only a limited range of colours; they melted on a hot day and broke on cold one; and they became rancid if not used for some time. In particular, they left embarrassing traces on whatever the wearer touched with her lips, such as cups, drinking glasses, cheeks, and cigarettes.

All these faults were eventually corrected thanks to cosmetic chemists. They researched the best waxes, oils, pigments, and skin protection agents, with the result that modern lipsticks should suffer none of the original failings. Modern lipsticks also contain fragrances to mask the greasy smell of the waxes. The following is a typical example.

### Lancôme® L'Absolu Rouge® Lipstick

*Contents*

Pentaerythrityl isostearate/caprate/caprylate/adipate
Macadamia ternifolia seed oil
Octyl dodecyl stearate
Microcrystalline wax
PEG-45/dodecylglycol copolymer
Polyglyceryl-3 beeswax
Ethyl hexyl methoxycinnamate
Hydrogenated castor oil dimer dilinoleate
Bisdiglyceryl polyacyl adipate
Alumina
Tin oxide
Geraniol
Calcium aluminium borosilicate
Calcium sodium borosilicate
Madecassoside
Methicone
Silica
Hydroxycitronellal
Hydroxypalmitoyl sphinganine
Hydroxypropyl tetrahydropyrantriol
Citronellol
Synthetic fluorphlogopite
Disteardimonium hectorite
Polyester-3
Polyethylene terephthalate
Polymethyl methacrylate
Methyl-2-octynoate
Hexylcinnamal
Benzyl alcohol
BHT
Tocopheryl acetate
Dye

## Emollients

**Pentaerythrityl isostearate/caprate/caprylate/adipate** consists of a mixture of interlinked fatty acids. It moisturises the lips, making them soft and smooth.

**Octyl dodecyl stearate** is a viscous liquid that slows down water loss from the lips, thereby making them appear plumper. It also absorbs damaging UV rays.

**Madecassoside** is a plant extract (*Centella asiatica*) that is said to have anti-ageing properties. It also counteracts inflammation.

**Hydroxypropyl tetrahydropyrantriol** can penetrate and strengthen the skin by increasing its connective tissue.

**Tocopheryl acetate** (also known as vitamin E) protects the lips because it can penetrate the skin where it also acts as an antioxidant.

## Waxes and Oils

**Macadamia ternifolia seed oil** has a silky feel and comes from the nut of the Australian tree *Macadamia integrifolia*.

**Microcrystalline wax** provides solid support.

**Polyglyceryl-3 beeswax** is chemically modified beeswax and is remarkably smooth.

**Bisdiglyceryl polyacyl adipate** is a grease based on natural oils and fats.

**Methicone** is silicone polymer and it protects the film of lipstick.

**Hydroxypalmitoyl sphinganine** (also known as ceramide) is a natural oil found in skin cells.

## Emulsifiers

**PEG-45/dodecylglycol copolymer** stabilises the mixture of the other ingredients.

**Hydrogenated castor oil dimer dilinoleate** ensures the dye molecules in the lipstick remain evenly dispersed and remain so when on the lips.

**Polyester-3, polyethylene terephthalate,** and **polymethyl methacrylate** are polymers that prevent the lipstick from cracking so that it remains a smooth film.

## Colourants

**Alumina** and **tin oxide** stabilise the colouring agents by bonding to them.

**Calcium aluminium borosilicate** and **calcium sodium borosilicate** are microparticles of glass added to make the lipstick shine.

**Dye** is what gives the lipstick its desired colour, in this case the dye is a deep rouge.

## Fragrances

Perfume is a complex mixture, some items of which have to be identified in a label because some people may be sensitive to them. **Geraniol** smells of roses, **hydroxycitronellal** smells of lime, **citronellol** smells of geraniums and roses, **methyl-2-octynoate** smells of violets, and **hexylcinnamal** smells of camomile.

**Ethyl hexyl methoxycinnamate** is a sunscreen chemical and protects the lips against UV.

**Silica** is silicon dioxide, **synthetic fluorphlogopite** is an aluminium silicate fluoride mineral, and they provide bulk.

**Disteardimonium hectorite** acts as thickener and gives the lipstick support.

**Benzyl alcohol** is a preservative.

**BHT** (butylated hydroxytoluene) is an antioxidant to protect the other ingredients.

Other popular brands of lipstick include L'Oreal®, Rimmel®, and Max Factor®. They also offer the same benefits as the one above and, like Lancôme® lipsticks, they also offer a remarkable range of shades, textures, and colours.

### 3.   MASCARA

*The Chemistry*

Mascara has been worn on the eyelashes and around the eyes since the days of the Ancient Egyptians. In those times they used kohl, which was made by grinding the black ore galena (lead sulfide) with oil or soft wax. Kohl is still used in parts of the world even today, but has been banned in others on account of the lead it contains.

Modern mascara relies on the black mineral iron oxide and this is applied in the form of a paste with the aid of a specially shaped brush. The user wants it to look good, be easy to apply to the lashes, and not irritate the eye or eyelid so it has to be free of microbes and possible irritants. Cosmetic chemists have come up with products that satisfy these needs and they have done this with a blend of waxes, oils, solvents, and pigments to get the texture just right. Although the number of ingredients in a typical mascara seems to be somewhat large, they can be divided into the categories just mentioned. Two examples are given, one a standard product, the other with something rather unusual added.

© Shutterstock

## L'Oreal® Volume Million Lashes® Extra-Black

*Contents*

Aqua/water
Paraffin
Potassium cetyl phosphate
Cera alba/beeswax
Cera carnauba/carnauba wax
*Acacia senegal/Acacia senegal* gum
Glycerin
Cetyl alcohol
Propylene glycol
Hydroxyethylcellulose
Phenoxyethanol
PEG/PPG 17/18 dimethicone
Steareth-20,
Phenethyl alcohol
Sodium polymethacrylate
Hydrogenated jojoba oil

Hydrogenated palm oil
Pentaerythrityl tetraisostearate
PVP
Disodium EDTA
Polyquaternium-10
Methyl paraben
Soluble collagen
Silica
Panthenol
Silica dimethyl silylate
Sodium chondroitin sulfate
Atelocollagen
CI 77266
CI 77491
CI 77492

*Solvents*

**Aqua/water** supports the pigments. (**Disodium EDTA** is a sequestering agent designed to prevent metal ions from interfering.)
**Paraffin** dissolves the oils.

*Emulsifiers*

**Potassium cetyl phosphate** and **steareth-20** help the waxes and oils to blend with the water-soluble ingredients.

*Wax Base*

**Cera alba/beeswax, cera carnauba/carnauba wax** and *Acacia senegal/Acacia senegal* **gum** (also known as gum Arabic) are blended to give the right consistency.

*Emollients*

These are **glycerine, cetyl alcohol, propylene glycol,** and **pentaerythrityl tetraisostearate**. The first of these prevents the mascara from drying out.
**Panthenol** is related to vitamin B5 and is thought to have beneficial effects on the skin.
**Sodium chondroitin sulfate** is a skin conditioner.
**Atelocollagen** is a protein that is normally used as a skin conditioner.

*Viscosity Control*

**Hydroxyethylcellulose, sodium polymethacrylate, PVP,** and **polyquaternium-10** are polymers designed to ensure that the mascara has the right consistency and remains in place once it has been applied.

**Soluble collagen** acts as a thickener.

*Antimicrobial Agents*

**Phenoxyethanol, phenylethyl alcohol,** and **methyl paraben** are antimicrobial agents needed to ensure the mascara is free of any germs that might cause eye infections.

*Oils*

**Hydrogenated jojoba oil** and **hydrogenated palm oil** are natural plant oils that have had their unsaturated bonds removed to prevent them going rancid.

**PEG/PPG 17/18 dimethicone** is silicone oil to which different polymers have been attached so it will blend better with the other ingredients.

*Fillers*

**Silica** and **silica dimethyl silylate** support the mascara to ensure it remains firm.

*Colourants*

**CI 77266** is carbon black, while **CI 77491** and **CI 77492** are black iron oxides.

**Rituals**®️ **Perfection**

For a special look there are products with additional ingredients. The following product actually contains ground-up gems.

*Contents*

| | |
|---|---|
| Aqua/water | Palmitic acid |
| *Copernicia cerifera* (Carnauba) wax | Candelilla cera/*Euforbia cerifera* wax |
| Paraffin | Triethanolamine |
| Synthetic beeswax | Oleic acid |
| Acacia Senegal gum | Stearic acid |
| Acrylates copolymer | Magnesium silicate |

PEG/PPG 17/18 dimethicone
Phenoxyethanol
Caprylyl glycol
Aminomethyl propanediol
Hydroxyethylcellulose
Sodium dehydroacetate
Potassium sorbate
Butylene glycol
Dimethicone
Glycerin

Arachidic acid
Behenic acid
Myristic acid
Amethyst powder
Sapphire powder
Ruby powder
*Chrysanthellum indicum* extract
*Alteromonas* ferment extract
CI 77499 Iron oxide

## Solvents

These are **aqua/water** and **paraffin,** which dissolve the various kinds of ingredients.

## Mascara Base

Mascara has to be waterproof and so consists mainly of oils and waxes.

*Copernicia cerifera* **(Carnauba) wax** and **Candelilla cera/ *Euforbia cerifera* wax are natural waxes. Synthetic beeswax is** manufactured

**Palmitic acid, oleic acid, stearic acid, arachidic acid, behenic acid,** and **myristic acid** are fatty acids that are all of plant origin. These blend in with the waxes to give the right consistency.

**Triethanolamine** and **aminomethyl propanediol** are buffering agents.

**PEG/PPG 17/18 dimethicone** and **dimethicone** are silicone oils.

## Viscosity Control

**Acacia Senegal gum** helps the mascara adhere to the eyelashes. **Hydroxyethylcellulose** and **butylene glycol** control the lubricity.

## Emulsifier

**Acrylates copolymer** acts as an emulsifier and helps the ingredients to blend well.

## Fillers

**Magnesium silicate**, also known as talc is a natural mineral renowned for its softness.

*Preservatives*

**Sodium dehydroacetate, potassium sorbate, phenoxyethanol** and **caprylyl glycol** are antimicrobial agents and preservatives.

*Emollients*

**Glycerin** stops the mascara from drying out.

***Chrysanthellum indicum* extract,** also known as golden chamomile, and ***Alteromonas* ferment extract** are anti-inflammatory agents believed to be good for the skin.

*Colourants*

**Amethyst powder, sapphire powder,** and **ruby powder** add a little sparkle to the mascara, and are made by grinding the artificially produced gemstones.

**CI 77499** is black iron oxide known as magnetite.

Other brands of mascara include Rimmel® and Max Factor®.

## 4.  NAIL POLISH

*The Chemistry*

If you paint your nails, you need something that coats them with a tough but flexible film that adheres strongly to the nail and comes in a choice of colours and shades. As with most paints, the colouring agent is dissolved in a suitable solvent which then evaporates, leaving a film of varnish behind.

The requirements for nail polish are that the solvent is harmless, the colour is the one you want, the painted nail lasts for several days and is not affected by other activities that involve the hands and that might chip it. The solvent is generally a simple ester of acetic acid (ethanoic acid) in which the ester group is a short hydrocarbon chain. The product also needs to contain emollients, preservatives, and of course intensely coloured dyes.

## L'Oréal® Paris Colour Riche Nails Femme Fatale

This is a classic red nail varnish.

*Contents*

Ethyl acetate
Butyl acetate
Nitrocellulose
Propyl acetate
Isopropyl alcohol
Tributyl citrate
Tosylamide/epoxy resin
Adipic acid/neopentyl glycol/
  trimellitic anhydride copolymer
Stearalkonium hectorite
Acrylates copolymer
Benzophenone-1
Polyacrylate-4
Acetyl tributyl citrate
Calcium sodium borosilicate
Synthetic fluorphlogopite

Alumina
Dimethicone
Polyethylene terephthalate
Calcium aluminium borosilicate
Citric acid
Phthalic anhydride/glycerin/
  glycidyl decanoate copolymer
Poly(1-hydroxy-1-phenyl-propylene)
Polyurethane-11
Oxidized polyethylene
CI 77002 Aluminium hydroxide
Red colorants:
CI 15880 Red 34 Lake,
CI 15850 Red 7 Lake,
CI 75470 Carmine.

*Solvents*

The main ones are **ethyl acetate, butyl acetate,** and **propyl acetate,** which dissolve the varnish and evaporate slowly.

A little **isopropyl alcohol** is also included to solubilise certain components, and it is also quite volatile.
**Dimethicone** is silicone oil.

*Varnishes*

**Nitrocellulose** will form a colourless layer, along with other film-forming chemicals, namely **tosylamide/epoxy resin, adipic acid/neopentyl glycol/trimellitic anhydride co-polymer, polyacrylate-4,** and **polyethylene terephthalate.** These have been blended to give just the right amount of flexibility and strength. Also needed are **tributyl citrate** and **acetyl tributyl citrate,** which help the polymers to spread more smoothly and evenly without drips.

**Poly(1-hydroxy-1-phenyl-propylene)** and **polyurethane-11** are also plastic film-formers. **Phthalic anhydride/glycerin/ glycidyl decanoate copolymer** makes the varnish more flexible.

*Emulsifier*

**Acrylates copolymer** is an emulsifier that keeps all the ingredients well mixed.

*Viscosity*

**Oxidized polyethylene** is used to get the right viscosity.

*Colourants and Texture*

**Stearalkonium hectorite** is based on the mineral hectorite, which is a magnesium lithium silicate. It provides a pearlescent sheen, as does **synthetic fluorphlogopite,** which is a potassium magnesium silicate mineral.

**Calcium sodium borosilicate** and **calcium aluminium borosilicate** give the varnish a glass-like texture.

The ingredients list contains around 15 colourants that are described as ' + / − may contain' and of which the following are identified as red: **CI 15880,** which is also known as **Red 34 Lake; CI 15850,** which is **Red 7 Lake;** and **CI 75470** is **carmine.**

**CI 77002** is **aluminium hydroxide** and acts as a mordant for the colourants, as does **alumina,** which is aluminium oxide.

(**Citric acid** reacts with the aluminium compounds to form aluminium citrate, which solubilises the aluminium.)

**Benzophenone-1 protects** the nail varnish from being affected by UV.

There are over 250 different nail varnishes to choose from, under brand names like Rimmel®, Max Factor®, Pixel®, and Essie®.

## 5.   ANTI-WRINKLE CREAM (ANTI-AGEING CREAM)

*The Chemistry*

There is a natural desire never to grow old and while it is not preventable at least it is possible to disguise the fact. The two things that signal oncoming old age are hair and skin. The former we dealt with under Chapter 3, The Bathroom. Here we look at the skin.

The skin is the largest organ of the body and it eventually reveals that we are ageing. Thankfully, most of it is hidden beneath clothing but the face is exposed and this creates a demand for ways to hide what it may reveal. There are products on the market that can partly repair the ravages of time. (And of course there are surgical methods of doing this.)

Cosmetic chemists have been trying to find remedies and not entirely without success. The ideal product would remove the top layer of dead skin, and there are substances that can do that. Then a product needs to tighten the layer beneath and make it look plumper, and again there are some to do this. It should also leave a layer on the surface that fills wrinkles and leaves it looking smooth. Finally the cream should protect the skin against moisture loss, against the effects of UV from the sun, and against urban pollution. Again there are ingredients that can achieve this to a certain extent.

Such products have been devised and they do provide benefits. The following one makes the claim it has been 'independently proven to restore skin's structure, visibly improving deep lines and wrinkles by up to 50%.' This product is a collection of ingredients that might make it possible

What an anti-ageing product needs are emollients to soften the skin, humectants to rehydrate it (and keep the cream moist), antioxidants to protect it, with preservatives and antimicrobial agents to ensure the cream remains germ-free, and emulsifiers to ensure all the ingredients blend nicely together.

### Boots No.7 Protect and Perfect® Intense Beauty Serum

*Contents*

Cyclopentasiloxane
Aqua
Butylene glycol
Dimethicone cross-polymer
Cyclohexasiloxane
Glycerin
Arabinogalactan
Sodium ascorbyl phosphate

Magnesium sulfate
Dimethicone copolyol
Phenoxyethanol
Sodium PCA
Cetyl PEG/PPG-10/1 dimethicone
Hexyl laurate
Polyglyceryl-4 isostearate
Retinyl plamitate
*Medicago sativa* (Alfalfa) extract
Methyl paraben
Propylene glycol
*Lupinus albus* seed extract

Butyl paraben
Ethyl paraben
Carbomer
Propyl paraben
Isobutyl paraben
Polysorbate 20
*Panax ginseng* root extract
*Morus alba* leaf extract
Tocopherol
Palmitoyl oligopeptide
Palmitoyl tetrapeptide-7

## Emollients

Those that are most effective are silicone oils, namely **cyclopentasiloxane, dimethicone cross-polymer, cyclohexasiloxane, dimethicone polyol,** and **cetyl PEG/PPG-10/1 dimethicone**. They make the skin feel smooth, they prevent moisture evaporating, and are long-lasting.
**Hexyl laurate** also softens and smoothes the skin.

## Humectants

These prevent loss of water from the skin, keeping it plumper looking. The ones in this product are **butylene glycol, glycerin, propylene glycol,** and **sodium PCA** (sodium pyrrolidone carbonic acid). The last of these is a humectant naturally present in skin.

## Anti-Ageing Ingredients

This product contains two substances that might have anti-ageing properties. These are **palmitoyl oligopeptide** and **palmitoyl tetrapeptide-7**. These consist of protein chains (also known as peptides) attached to the long chain fatty acid palmitic acid, which comes from palm oil. They increase skin thickness and thereby help to hide wrinkles.
**Magnesium sulfate** provides magnesium ions, which are thought to be beneficial to the skin.

## Plant Extracts

Plants contain dozens of chemicals, some of which might also be of benefit to the skin, although the evidence is often either based on folklore or alternative lifestyle PR. The ones in this product are as follows:

**Arabinogalactan** is a carbohydrate polymer that might help tighten the skin by allowing skin cells to adhere to one another.

*Medicago sativa* is extracted from the alfalfa plant and is reputed to be anti-ageing, although there is no scientific basis for this.

*Lupinus albus* **seed extract** comes from lupin seeds and is reputed to stimulate the production of collagen, the fibrous protein that gives skin its texture and the lack of which makes skin sag and wrinkle.

*Panax ginseng* **root extract** is more generally taken as a drink and is supposed to have 'rejuvenating' health benefits.

*Morus alba* **leaf extract** comes from mulberry leaves and is said to improve skin tone.

*Antioxidants*

These are simply more of the natural substances that our body needs for many purposes, namely **sodium ascorbyl phosphate** (a derivative of vitamin C), **retinyl palmitate** (also known as vitamin A), and **tocopherol** (also known as vitamin E). These neutralise the free radicals, produced by UV and oxygen, which are chiefly responsible for skin ageing.

*Emulsifiers*

These are **polyglyceryl-4 isostearate**, **carbomer**, and **polysorbate 20**, which are needed to get the various ingredients to blend smoothly together.

*Antimicrobials*

**Methyl, ethyl, propyl, butyl** and **isobutyl parabens** are there to prevent bacteria breeding in the face cream and in the various ingredients from which it is made.

There are over 130 anti-wrinkle, anti-ageing creams and lotions on the market in the UK, and these are sold under brand names such as Dove®, Olay®, Nivea®, and L'Oréal®. Many offer the same benefits as the one above.

## 6. PERFUME

*The Chemistry*

The use of perfume is now extremely popular, and there are hundreds of such products to choose from. The ingredients of a perfume are fragrance molecules, solvents, preservatives, moisturisers, and colours.

Perfumes are made up of fragrance chemicals, most of which have been produced by chemists and these are to be preferred to those produced naturally because they can be absolutely pure and free from possible contaminants of the kind that cause natural fragrances to change over time. Once the molecular structure of a natural fragrance has been deduced then it can be made in the laboratory and maybe even modified slightly to produce a different note.

Composing a perfume is analogous to composing music and the terms are similar. High notes are ones that are most volatile, evaporate quickly, and convey a fresh feeling. Middle notes are less volatile and are often the very heady perfume of oriental flowers. Base notes are the least volatile of all and may have an animalistic smell. Current trends, however, are for lighter note perfumes and these have fewer base notes of the kind that were once taken from animals like the civet cat and the musk deer, although even the molecules associated with those smells can now be synthesised.

To compose a new perfume you need a blend of many chemicals to achieve a unique fragrance, and the skill of the perfume chemist is to find a combination that evaporates together so that no smell dominates. Moreover, the perfume must continue to smell the same throughout the day and into the evening. This requires the right combination of molecules to be blended. When liquids combine, the rate of evaporation of the resulting mixture will be different from that of their individual components, and yet all must evaporate together to achieve the desired effect. This requires specialist chemical knowledge.

The usual solvent for perfumes is alcohol and the final solution consists of 15% of fragrance ingredients dissolved in this solvent. There are more dilute versions, such as eau de toilette, and these are priced accordingly.

## Insolence® Eau de Parfum by Guerlain® of Paris

This was launched in 2008 and is typical of the kind of perfume women now prefer.

*Contents*

Alcohol
Parfum
Aqua
Alpha-isomethyl ionone
Linalool
Ethylhexyl methoxycinnamate
Limonene
Coumarin
BHT
Butyl methoxydibenzoylmethane
Butylene glycol dicaprylate/dicaprate
Geraniol
Citronellol
Isoeugenol

Citral
Citric acid
Methyl 2-octynoate
Benzyl benzoate
Benzyl salicylate
Farnesol
Benzyl alcohol
Eugenol
Disodium EDTA
Hydroxycitronellal
CI 60730 (violet)
CI 14700 (red)
CI 19140 (yellow)

*Solvents*

These are **alcohol**, which is ethanol and **aqua**, which is water.

*Fragrances*

**Parfum** here refers to fragrances of which some, in theory, may be secret or at least their relative amounts are. Others have to be identified because some individuals may be sensitive to them and these are **alpha-isomethyl ionone**, which is described as having a 'clean' floral smell. **Linalool** has a floral smell with a hint of spice. **Limonene** is found in the peel of oranges. **Coumarin** has the scent of new-mown hay. **Geraniol** smells of roses. **Citronellol** is the main fragrance of geraniums and roses. **Isoeugenol** smells of vanilla. **Citral** smells of lemon. **Methyl 2-octynoate** smells of violets. **Eugenol** has a spicy smell like cloves. **Hydroxycitronellal** smells of lime. **Farnesol** enhances the floral smell of the various perfumes.

*Preservatives*

These are **ethylhexyl methoxycinnamate** and **butyl methoxydibenzoylmethane,** which act as UV absorbers to protect the other ingredients.

**BHT** (butylated hydroxytoluene), **benzyl benzoate, benzyl salicylate**, and **benzyl alcohol** are preservatives for the various fragrances. Benzyl salicylate also helps the fragrance molecules to blend together.

**Citric acid** makes the product slightly acidic and this helps preserve the fragrance molecules, which would be affected by alkaline conditions.

**Disodium EDTA** is there to sequester any metal ions that might interfere with the other ingredients.

*Moisturiser*

This is **butylene glycol dicaprylate/dicaprate,** which soothes the skin against any irritation that might be caused by the fragrance molecules.

*Colours*

**CI 60730** is violet, **CI 14700** is red, and **CI 19140** is yellow.

Many perfumes come from perfume houses that have been long established. The above one comes from Guerlain®. Other famous ones are Estée Lauder® and Givenchy®, but there are many more.

## 7.  BODY CREAM

*The Chemistry*

When you shower every day you are washing off protective body oils, which you then may feel the need to replace with emollients (moisturisers). Some will act as a barrier, preventing moisture loss from the skin so it won't feel dry and itchy, others might even be absorbed by the skin and benefit it that way, by relaxing it and softening it. This type of product will need to consist of emollients, humectants, natural oils, emulsifiers, fragrances, antioxidants, and preservatives.

### Rituals® Touch of Happiness

This product appears to have been formulated to include all the chemicals that are supposed to be good for the skin.

*Contents*

Aqua/water
Cyclopentsiloxane
Cetearyl alcohol
Glycerin
Caprylic/capric triglyceride
Parfum
Ceteareth-20
Phenoxyethanol
*Helianthus annuus* (sunflower)
   hybrid oil
*Centena asiatica* (gotu cola)
   leaf extract
*Juniperus virginiana*
   (cedar) wood extract
*Bambusa vulgaris* leaf extract
*Citrus aurantium dulcis*
   (orange fruit) extract
Dimethicone
Propylene glycol
Glyceryl stearate

Isohexadecane
Tocopheryl acetate
Ceteareth-12
Carbomer
Caprylyl glycol
Allantoin
Tocopherol
Potassium sorbate
Linalool
Cetyl palmitate
Alcohol
Olus (vegetable) oil
Limonene
Sodium hydroxide
Hexyl cinnamal
*Helianthus annuus* (sunflower)
   seed oil
Butylene glycol
Citronellol
Ubiquinone

   **Aqua/water** is a solvent.

*Emollients*

   **Cyclopentsiloxane** and **dimethicone** are silicone oils. The first of these has the fluidity of water, making it easy to spread over the skin, whereas the dimethicone gives a permanent protective layer, making the skin feel smooth and helping it retain moisture.

**Cetearyl alcohol** acts as a thickener as well as being an emollient.

**Glyceryl stearate** acts as a skin conditioning agent as well as being an emollient.

**Isohexadecane** is a branched hydrocarbon chain liquid with emollient properties.

**Allantoin** is said to rehydrate living skin cells.

**Cetyl palmitate** is an emollient and skin conditioner.

**Butylene glycol** is naturally present in skin and prevents loss of water from the skin.

*Natural Oils*

These contain dozens of chemicals, some of which might also be of benefit, although the evidence is often either based on folklore or alternative lifestyle PR. The ones in this product are as follows:

**Caprylic/capric triglyceride**, which comes from coconut oil.

*Helianthus annuus* **(sunflower) hybrid oil** and *Helianthus annuus* **(sunflower) seed oil** are grown from different seed types, with the former giving greater yields.

*Centena asiatica* **(gotu cola) leaf extract** contains saponins, which are thought to be good for the skin. These compounds were used as surfactants before soap was discovered.

*Juniperus virginiana* **(cedar) wood extract** has long been used as a herbal remedy to treat various conditions, and it has a pleasant balsamic smell.

*Bambusa vulgaris* **leaf extract** is also known as the golden bamboo and is used in herbal medicine.

*Citrus aurantium dulcis* **(orange fruit) extract** is used in herbal medicine.

**Olus (vegetable) oil**, which is an unspecified vegetable oil.

*Humectants*

**Glycerin** is the humectant and is there to keep the cream from drying out.

**Parfum** is a mixture of fragrances, some natural, some artificial, and it may contain items that some individuals are sensitive to. These are identified as **linalool**, which has a floral spicy note; **limonene,** which smells of oranges; **hexyl cinnamal,** which smells of camomile; and **citronellol**, which smells of geraniums and roses.

*Emulsifiers*

**Ceteareth-20** and **ceteareth-12** act as emulsifiers to blend all the ingredients together.

*Preservatives*

**Phenoxyethanol** and **potassium sorbate** prevent microbial contamination of the product. **Caprylyl glycol** is an antimicrobial agent. **Sodium hydroxide** is an alkali and as such will maintain the pH of the cream above 7 and protect other ingredients.

*Solvents*

These are **propylene glycol** and **alcohol.**

*Antioxidants*

**Tocopherol**, which is vitamin E, and **tocopheryl acetate**, which is a form of vitamin E that is able to penetrate the surface of the skin better, are antioxidants. **Ubiquinone** is both an antioxidant and skin conditioner.

*Viscosity improver*

**Carbomer**, which is a polymer of acrylic acid cross-linked with sugar molecules, acts as a viscosity improver.

Other products of this kind are to be found under the brand names E45®, Nivea®, Dove®, Vaseline®, *etc.*

## 8. HAIR PRODUCTS

### Hairspray

*The Chemistry*

Hair is mainly made of a protein. As such, it will grow naturally in a way determined by our genes with respect to colour and structure. When we want to change how it looks then we may spend a lot of time and money styling it and, when we have it looking just the way we want it, we need a fixative so that it remains that way. The outer layer of hair can become slightly positively charged and then we find it difficult to control so we need something that will remove static. This is more of a problem for women, more because their hair styles tend to be of longer hair.

The chemical solution to the problem is to spray the hair with a polymer that will coat the strands of hair with a film that will hold it in place and act as an antistatic agent, and yet be easily removed when we next wash our hair. We also will want it to be easy to comb and so the hairspray should contain a lubricating oil as well.

### Silvikrin®

*Contents*

| | |
|---|---|
| Alcohol denat. | Parfum |
| Butane | Linalool |
| Propane | Triethyl citrate |
| Isobutane | Benzyl salicylate |
| Octylacrylamide/acrylates/ butylaminoethyl methacrylate copolymer | Butylphenyl methylpropional |
| | Limonene |
| Aqua | Hexyl cinnamal |
| Aminomethyl propanol | Geraniol |
| PEG-12 dimethicone | Alpha-isomethyl ionone |
| | Citronellol |

*Solvents*

**Alcohol denat.** is denatured alcohol that contains a bitter-tasting chemical to deter its theft and misuse. (By denaturing the alcohol, the company also avoids paying excise duty.) Here it acts as the solvent for the various molecules.

**Aqua** is water and a solvent.

*Gases*

**Butane, propane**, and **isobutane** are the hydrocarbon gases that act as propellants. These are under pressure in the canister and they force out the contents through a nozzle when the pressure is released.

**Octylacrylamide/acrylates/butylaminoethyl methacrylate copolymer** are the polymers that hold the hair in place and also act as an antistatic layer.

**Aminomethyl propanol** is there to stabilise the pH of the hairspray.

**PEG-12 dimethicone** is solubilised silicone oil that makes it easier to brush and comb the hair.

**Parfum** is a mixture of fragrances, some natural, some artificial, and it may contain items that some individuals are sensitive to. These are **linalool**, which has a floral note, as has **butylphenyl methylpropional**, while **limonene** smells of oranges, **hexyl cinnamal** smells of camomile, **geraniol** smells of roses, **alpha-isomethyl ionone** smells of the open air, and **citronellol** smells of geraniums and roses.

**Triethyl citrate** acts as an antioxidant to protect the various fragrances.

**Benzyl salicylate** is a fixative – in other words it is there to help the fragrance molecules to blend in with the other ingredients. Of itself, it has almost no odour.

## L'Oréal® Elnett® Hairspray

This hairspray claims to have extra strength and to brush out easily.

*Contents*

| | |
|---|---|
| Alcohol denat. | Dimethicone copolyol |
| Dimethyl ether | Limonene |
| VA/Vinyl butyl benzoate/crotonates copolymer | Benzyl benzoate |
| | Alpha-isomethyl ionone |
| Cyclopentasiloxane | Geraniol |
| Hydroxycitronellal | Citronellol |
| *Everina prunastri*/oakmoss extract | Coumarin |
| PPG-3 methyl ether | Amyl cinnamal |
| Ethylhexyl methoxycinnamate | Parfum/fragrance |
| Aminomethyl propanol | |

*Solvent*

**Alcohol denat.** is denatured alcohol that contains a bitter-tasting chemical to deter its theft and misuse. (By denaturing the alcohol, the company also avoids paying excise duty.) Here it acts as the solvent.

*Propellant*

**Dimethyl ether** is the gas that acts as propellant. It is under pressure in the canister and forces out the content through a nozzle when the pressure is released.

*Polymers*

**VA/Vinyl butyl benzoate/crotonates copolymer** coats the hair to hold it in place and prevent static.

**PPG-3 methyl ether** is polypropylene glycol polymer modified with methyl groups, and **dimethicone copolyol** is a modified silicone oil with polymer strands attached. These coat the hair, holding it in place and restoring its natural gloss.

*Fragrances*

**Parfum** is a mixture of fragrances, some natural, some artificial, and it may contain items that some individuals are sensitive to. **Hydroxycitronellal** smells of lime, **limonene** smells of oranges, **alpha-isomethyl ionone** has an open air smell, **geraniol** smells of roses, **citronellol** smells of geraniums and roses, **coumarin** smells of new-mown hay, and **amyl cinnamal** smells like jasmine.

*Everina prunastri*/**oakmoss extract** is extracted from lichen and acts as a fixative for the other fragrances as well as providing a base note to complement them.

**Cyclopentasiloxane** is silicone and it acts as a lubricant to make the hair easy to comb.

**Ethylhexyl methoxycinnamate** acts to filter out UV rays to protect the spray on the hair.

**Aminomethyl propanol** is a buffering agent to keep the pH stable.

**Benzyl benzoate** is a preservative.

There are more than 100 kinds of hairspray, many of which are different versions of popular brands, including L'Oréal®, Pantene®, and TRESemmé®.

## Brylcreem®

Another way to hold hair in place is to use a cream or gel. One popular cream for men is Brylcreem®.

*Contents*

Aqua
Alcohol denat.
Acrylates/C1–2 succinates/
  hydroxyacrylates
copolymer
Panthenol
Parfum
Sorbitol
Propylene glycol
Acrylates/Beheneth-25 methacrylate
  copolymer
Triethanolamine
PPG-1/PEG-9 lauryl
  glycol ether
Polyether-1
Phenethyl alcohol
PPG-2 methyl ether
Sodium polyacrylate
Tetrasodium EDTA
Benzoic acid
Methylisothiazolinone

*Solvents*

**Aqua**, water, and **alcohol denat.**, ethanol, are solvents for the other ingredients. (Denat. means denatured and it includes a bitter-tasting chemical to deter its theft and misuse.)

**Tetrasodium EDTA** counteracts hard water by sequestering calcium ions that would otherwise interfere with the emulsion.

*Polymers*

**Acrylates/C1–2 succinates/hydroxyacrylates copolymer** are the stiffening agents that hold the hair in place.

**PPG-2 methyl ether** is polypropylene glycol polymer modified with methyl groups.

*Emollient*

**Panthenol** is related to vitamin B5 and is a precursor to this vitamin. It is supposed to have beneficial effects on the scalp.

*Humectant*

**Sorbitol, propylene glycol,** and **sodium polyacrylate** stop the gel from drying out when on the hair.

*Viscosity Control*

**Acrylates/Beheneth-25 methacrylate copolymer** and **polyether-1** control the viscosity and thickness of the gel.

*Emulsifiers*

**Triethanolamine** and **PPG-1/PEG-9 lauryl glycol ether** act as emulsifiers.

*Preservatives*

**Phenethyl alcohol, benzoic acid,** and **methylisothiazolinone** are preservatives and antimicrobials.
**Parfum** refers to unspecified fragrances.

Other brands of hair gel include Shock Waves® and VO5®.

## Hair Restorer

*The Chemistry*

As men get older their testosterone begins to accumulate in hair follicles, eventually blocking some of them so the hair stops growing. There appeared to be no cure for this condition beyond grafting new skin with hair on to the top of scalp, as was once done. Then a rather unexpected side effect of a new drug was discovered.

This drug had been designed to lower blood pressure by re-laxing arteries so it could flow more easily. Those on whom it was tested reported a side effect that they hadn't expected: their hair started growing again. What appeared to be happening was that increased blood flow to the scalp was assisting in the removal of the testosterone that was blocking the hair follicles. Clearly there was a market for this drug among men who were going bald. The new drug was marketed as Regaine® – not to be taken as a prescribed medical treatment, but to be applied externally to the hair as a way of achieving the desired effect. In this way it could be sold as an over-the-counter product.

## Regaine® Foam Extra Strength

*Contents*

| | |
|---|---|
| Minoxidil (5%) | Stearyl alcohol |
| Alcohol | Polysorbate 60 |
| Water | Butyl hydroxytoluene |
| Glycerol | SD alcohol 40-B |
| Cetyl alcohol | Propellants: propane, |
| Citric acid | butane and isobutane |
| Lactic acid | Fragrance |

*Active Agent*

**Minoxidil** makes up 5% of the solution, which has to be directly applied to the scalp every day. For most men, but not all, it does have the desired effect of renewing hair growth.

*Solvents*

**Alcohol, water,** and **SD alcohol 40-B** act as solvents, with the latter being industrial alcohol that has been denatured and made undrinkable by the addition of the bittering agent Bitrex®.

*Humectants*

**Glycerol** helps to increase contact of the product with the skin of the scalp.

**Citric acid** and **lactic acid** keep the pH constant and below pH 7. Lactic acid is a natural component of skin.

*Emulsifiers*

**Cetyl alcohol** and **polysorbate 60** act as emulsifiers.

*Preservative*

**Butyl hydroxytoluene** is a powerful antioxidant preservative.

*Propellant Gases*

**Propellants: propane, butane and isobutane** are the hydro-carbon gases used to eject the Regaine from the can. These are under pressure in the canister and they force out the content through a nozzle when the pressure is released.

**Fragrance** is not specified so does not contain ones that need to be listed as potential irritants for some people.

## 9.  SEX

*The Chemistry*

I have not included the contraceptive pill in *Chemistry at Home* because it is really a medical product and cannot be bought over-the-counter. However, I can include the morning-after pill because this can be obtained from a pharmacist, although a reason must be given for wanting it.

Once an egg has been fertilized, it will implant itself in the wall of the womb and that procedure may take two or three days, which gives time to prevent it happening. One way is to take a dose of something that triggers menstruation such as levonorgestrel or mifepristone, which are similar to the natural hormones that perform this function every month.

## Levonelle®

*Contents*

Levonorgestrel
Potato starch
Maize starch
Colloidal silica anhydrous

Magnesium stearate
Talc
Lactose monohydrate

*Active Agent*

**Levonorgestrel.** Each tablet contains 1.5 mg of this active agent which behaves like the normal hormones that cause menstruation.

**Potato starch** and **maize starch** ensure the pills quickly disintegrate in the stomach so the levonorgestrel can be absorbed.

**Lactose monohydrate** also acts as an excipient.

**Colloidal silica anhydrous** and **talc** bulk out the pills.

**Magnesium stearate** is added to the mixture from which the pills are stamped out and it prevents the material sticking to the stamping machinery.

Other morning-after pills are available, such as Mifegyne® (based on mifepristone) and Plan B® (also based on levonorgestrel).

## Condoms

*The Chemistry*

There are still some traditional condoms that were made from a part of a sheep's intestine, which is like a condom, and were the only ones available in the UK for hundreds of years. They are still available in Italy and the USA.

Manufactured condoms were originally made from rubber latex but polyurethane is now the preferred polymer because it cannot cause an allergic reaction. Condoms come in several shapes and are generally lubricated with silicone oil rather than an oil-based grease. They can also be ribbed or studded to increase sensation. The most popular brand is Durex® and its makers, Reckitt Benckiser®, offer a range of condoms including one that incorporates a gel on the inside of the tip. This contains nitro-glycerine, which is designed to stimulate the flow of blood to the penis thereby making it larger and maybe even compensate for an erectile dysfunction, although in this respect it cannot compete with the next product, Viagra®.

## Viagra®

*The Chemistry*

Viagra® is used to treat impotence or erectile dysfunction in men. An erection is caused naturally by the release of nitric oxide (NO) molecules in response to sexual stimulation. These molecules trigger an enzyme to release cyclic *guanosine monophosphate* (also known as cGMP), which increases blood flow to the penis and it becomes erect. Meanwhile another enzyme, *phosphodiesterase-5*, is actively removing cGMP, albeit not at a rate to cause the erection to subside – although in some men this enzyme's role can result in partial or unsustainable erections, hence the need for Viagra®. The molecule sildenafil deactivates *phosphodiesterase-5* so the penis stays more erect and for longer.

*Contents*

| | |
|---|---|
| Sildenafil citrate | Hypromellose |
| Microcrystalline cellulose | Titanium dioxide |
| Anhydrous dibasic calcium phosphate | Lactose |
| Croscarmellose sodium | Triacetin |
| Magnesium stearate | FD&C Blue aluminium lake |

*Active Agent*

**Sildenafil citrate** blocks the enzyme *phosphodiesterase-5*, although natural processes eventually release this again.

*Excipient*

**Microcrystalline cellulose** acts as an excipient, and helps the tablet to break up when it reaches the stomach.

**Croscarmellose sodium** is another form of cellulose but its chains are cross-linked. This too is an excipient, in this case causing the tablet to swell when it becomes moist.

**Hypromellose** acts partly as glue and partly as a controlled release agent for the sildenafil so that its effects will last longer.

**Triacetin** prevents the tablet from drying out and becoming powdery.

*Filler*

**Anhydrous dibasic calcium phosphate** and **lactose** are there to bulk out the tablet.

*Colour*

**FD&C Blue aluminium lake** is bright blue. A lake consists of a dye molecule bonded to a metal atom, the object being to make the dye insoluble. This means it will pass through the gut without being absorbed. **Titanium dioxide** is a brilliant white pigment that is added to make the tablet a paler shade of blue.

**Magnesium stearate** is a grease that prevents the ingredients from sticking to the machinery that stamps them into tablets.

Other products that produce prolonged erections include Levitra® and Cialis®.

**Personal Lubricant**

*The Chemistry*

Doctors and nurses may need to lubricate their gloved fingers or equipment when they are investigating the anus or vagina of a patient. A lubricant can also be used to supplement the natural secretions that the body produces when sexually aroused.

## K-Y® Jelly

This is the most popular brand of personal lubricant.

*Contents*

| | |
|---|---|
| Aqua | Benzoic acid |
| Propylene glycol | Polysorbate 60 |
| Sorbitol | Tocopheryl acetate |
| Hydroxyethylcellulose | |

**Aqua** is water and a solvent.

**Propylene glycol** and **sorbitol** act as lubricants and are compatible with water, unlike grease-based lubricants.

**Hydroxyethylcellulose** provides just the right viscosity and lubricity.

**Benzoic acid** is a powerful antimicrobial agent.

**Polysorbate 60** is an emulsifier.

**Tocopheryl acetate** is a form of vitamin E, and it can penetrate skin where it acts as a protective antioxidant.

There are other sexual lubricants with added extras – some to delay ejaculation by numbing the penis with the mild anaesthetic lidocaine, and some that offer some protection against conception and sexually transmitted diseases.

Other products are sold under the brand names of Aquagel®, Galpharm®, and others.

# CHAPTER 8

# The Kitchen

© Shutterstock

Chemistry is supposed to have its roots in cooking, and indeed cooking does effect chemical changes. Perhaps understandably, the modern kitchen is less of a chemical laboratory than it used to be. More and more we are eating prepared meals, takeaways, or going out to dine, and when we do cook at home we want to

Chemistry at Home: Exploring the Ingredients in Everyday Products
By John Emsley
© John Emsley, 2015
Published by the Royal Society of Chemistry, www.rsc.org

use fresh or frozen food, in the form of meat, chicken, fish, vegetables and fruit, combined with flour, sugar and butter, – none of which come with a list of ingredients. And while these may undergo complex chemical changes as we cook them, these are not something that will deter us from enjoying them – unless of course we burn things, and then we may produce molecules it would be best to avoid.

So what products are likely to be found in a kitchen and come with a list of components? Surprisingly few, unless the family consume lots of processed foods. Here we are concerned with items a cook might use. We will also look at some cleaning products, and two products that we might never need: one designed to ignite a flame, the other to detect flames that are a danger; in other words, a box of matches and a smoke detector.

The products in this chapter, and the brands analysed, are as follows:

1. Salt (Cerebos® Iodised Salt, LoSalt®)
2. Flavour enhancers (Kikkoman® Soy Sauce, Lea & Perrins® Worcestershire sauce)
3. Baking powder (Borwick's® Baking Powder)
4. Yeast extract (Marmite®)
5. Mayonnaise (Hellmann's® Light Mayonnaise)
6. Stock (Knorr® Chicken Stock Pot)
7. Gravy granules (Bisto® Best Rich Roast Beef Gravy)
8. Washing-up liquid (Fairy® Liquid Platinum)
9. Dishwasher tablets (Finish® Quantum®)
10. Oven cleaner (Mr Muscle® Oven Cleaner)
11. Matches (The Original Cook's® Matches)
12. Smoke alarm
13. Kettle and iron descaler (Ecozone® Kettle and Iron Descaler)

## 1. SALT

*The Chemistry*

It is common knowledge that salt is sodium chloride. It reaches us in two forms depending on its source: it can be salt from a mine or salt from the sea. In fact, the former is salt from an ancient dried-up sea whereas the latter has been obtained from the oceans of today. In both cases the product has to be made suitable for human consumption, which means it has to be dissolved in clean water and recrystallised to remove impurities. Table salt may also have an anti-caking chemical added to it, such as sodium aluminosilicate, magnesium carbonate, or sodium hexacyanoferrate. These ensure it remains free running. Tiny salt crystals by themselves tend to stick together.

Salt can also have something added to it: iodine. This is a kind of salt that the United Nations is promoting as having global implications for health. In certain parts of the world, like China and India, the soil is deficient in iodine and plants that grow in such regions cannot provide enough of this essential element for the human diet. The average person has between 10 and 20 milligrams of iodine in their body; only tiny amounts are needed on a daily basis, and iodised salt can provide it.

© Shutterstock

Iodine is needed by the thyroid gland to make thyroxine and tri-iodothyronine, molecules that have four and three iodine atoms, respectively. These hormones regulate several metabolic functions, in particular the body's temperature, so are needed throughout life. They are also vital for the development of the brain in the unborn baby. Generally, a balanced diet will provide all the iodine we need but not always.

In other respects, salt is something we should moderate because a high sodium level in the blood raises blood pressure. Most processed food manufacturers have significantly reduced the amount they add to their products, or found alternative flavours, such as soy sauce – see section 2 of this chapter.

### Cerebos® Iodised Salt

*Ingredients*

Salt                              Anti-caking agent: sodium
Potassium iodate         hexacyanoferrate(ɪɪ)

**Salt** is sodium chloride.

**Potassium iodate** has the chemical formula $KIO_3$ and is added in preference to potassium iodide (KI). The latter has a tendency to be oxidised to iodine under damp conditions and the iodine can then evaporate.

**Sodium hexacyanoferrate** has the food code E535 and is only approved as a food additive for use in salt and salt substitutes. It changes the shape of salt crystals so that they don't stick together and become lumpy.

Another common brand of table salt is Saxa®. There are also sea salts, such as Maldon® Sea Salt, Natural Halen Môn Anglesey Sea Salt®, and many more. All are essentially the same chemical although they may contain minute traces of other minerals that are to be found in the sea.

### Salt Substitute

*The Chemistry*

The central nervous system passes messages to and from the brain by means of sodium and potassium ions moving in and out of nerve cells, which is equivalent to a flow of current along them. We could not live without taking in sodium and potassium in our diet because these elements are continually lost in our urine and our sweat. We have 100 grams of sodium in our body and lose around 3 grams of this a day; we have around 120 grams of potassium and lose around 3.5 grams a day. These amounts are replaced naturally by the foods we eat and so we rarely need to add more unless we engage in strenuous physical activity.

Salt does bring out flavour and so many people continue to add it to vegetables or sprinkle it on the food on our plates; so the average daily intake is more like 6 grams a day or more.

If you suffer from high blood pressure your doctor may tell you to reduce your sodium intake. This can be done without much difficulty, but sometimes it might be necessary to use a substitute to replace the sodium in a recipe with the element that most resembles it, and that is potassium. These elements are neighbours in the same group of the periodic table.

## LoSalt®

*Ingredients*

| | |
|---|---|
| Potassium chloride 66% | Magnesium carbonate |
| Sodium chloride 33% | Hexacyanoferrate |

**Potassium chloride 66%** is the substitute salt, KCl.

**Sodium chloride 33%** is ordinary salt, NaCl.

**Magnesium carbonate** helps keep the salt free of moisture and prevent caking.

**Sodium hexacyanoferrate** has the food code E535 and is only approved as a food additive for use in salt and salt substitutes. It changes the shape of salt crystals so that they don't stick together and become lumpy.

LoSalt® also comes as an iodised version, but is it not available in the UK as yet. The iodised version includes potassium iodide.

## 2.  FLAVOUR ENHANCERS

The two products described in this section are available in the kitchen as a way of enhancing the savoury flavour of a dish, which they do by boosting its umami component. Umami is the fifth kind of taste sensation and has only recently been recognised as such in the UK. It now takes its place with the other four basic tastes: salt, acid (also known as sour), sweet, and bitter. Umami was first recognised as a basic taste by Kikunae Ikeda in Japan in 1908.

### The Chemistry

Soy sauce has long been a part of oriental cooking because it enhances the umami or savoury flavour of dishes. It can be made by a traditional fermentation process, which is slow, or it can be made quicker by the acid hydrolysis of soya protein. What it provides is monosodium glutamate (MSG), which people once thought was the chemical that caused a condition called Chinese Restaurant Syndrome. There is no such medical condition, as later research proved, although it is possible to consume more MSG than the body can cope with and then the face becomes flushed and you suffer a headache. These symptoms will disappear within 24 hours as the excess MSG is absorbed or excreted. In any case, MSG is a natural part of every living cell so you cannot be allergic to it, only intolerant of taking in too much at one meal.

The following product is made by the traditional method.

### Kikkoman® Soy Sauce

#### Ingredients

| | |
|---|---|
| Monosodium glutamate | Sodium chloride |
| Glucose | Lactic acid |
| Fructose | Succinic acid |
| Leucine | Acetic acid |
| Lysine | Alcohol |
| Potassium chloride | |

**Monosodium glutamate** (MSG) is the basis of the umami flavour.

**Glucose** is a natural carbohydrate.

**Fructose** is the carbohydrate that provides the sweetness of honey and treacle, and gram-for-gram it is much sweeter than sugar.

**Leucine** is an essential amino acid. In other words it has to be part of our diet since we cannot synthesise it within the body, as we can for most other amino acids. It is present in cereal proteins.

**Lysine** is also an essential amino acid and it is present in red meat, eggs, beans, and fish.

**Potassium chloride** and **sodium chloride** are essential for the working of the nervous system.

**Lactic acid** is also known as milk acid. This is produced naturally in the human body in large amounts as a by-product of the metabolism of carbohydrates.

**Succinic acid** has the food code E363 while **acetic acid** is E260. These regulate the acidity of the sauce. They also act as natural preservatives.

**Alcohol** is a by-product of fermentation.

Other brands of soy sauce are available, such as Sharwood's®, Amoy®, *etc.*

## Lea & Perrins® Worcestershire Sauce

*The Chemistry*

This traditional flavouring agent is added to tomato juice, soups, and savoury dishes; its tang is unmistakable. It is free of artificial colours and preservatives, so what preserves it during its many weeks on a kitchen shelf? The answer is its acetic acid content, which comes from its malt vinegar and spirit vinegar.

*Ingredients*

| | |
|---|---|
| Malt vinegar (from barley) | Tamarind extract |
| Spirit vinegar | Onions |
| Molasses | Garlic |
| Sugar | Spice |
| Salt | Flavouring |
| Anchovies | |

**Malt vinegar (from barley)** and **spirit vinegar** ensure that the pH of this product is 3.8, and this provides a medium in which microbes cannot flourish.

**Molasses** and **sugar** provide a little sweetness.

**Salt** is sodium chloride.

**Anchovies** provide the umami flavour.

**Tamarind extract** has a unique sweet sour flavour. It is the same flavouring that is also found in HP® sauce.

**Onions** and **garlic** provide sulfoxides with their unique kinds of flavours.

**Spice** is not specified.

**Flavouring** is not specified.

### 3. BAKING POWDER

*The Chemistry*

Baking powders have been around for more than 100 years and were designed to make confectionary lighter. They do what yeast does naturally, but very slowly, when making bread. Baking powder generates bubbles of carbon dioxide that remain trapped in the dough. This makes products like cakes lighter in texture.

The chemical reaction that produces $CO_2$ is an acid attack on a carbonate, which comes in the form of sodium hydrogen carbonate, $NaHCO_3$, also known as sodium bicarbonate. Until the acid and carbonate come together in the presence of water, and react, they need to be kept apart as solids, and while this is not a problem for carbonates, which are all solids, it is for acids, which are often liquids. What is needed is a solid acid salt, and the ones used are calcium hydrogen phosphate, $CaHPO_4$, or disodium dihydrogen diphosphate $Na_2H_2P_2O_7$. Some older baking powders used 'cream of tartar' as the source of acid and this is mono-potassium tartrate, $KC_4H_5O_6$.

A typical baking powder is as follows:

#### Borwick's Baking Powder®

*Ingredients*

Wheat flour
Acid sodium pyrophosphate
Sodium bicarbonate

**Wheat flour** is just what it says. (There is also a cornflour version for those susceptible to gluten.)

**Acid sodium pyrophosphate** is disodium dihydrogen diphosphate, $Na_2H_2P_2O_7$, and it has the food code E450.

**Sodium bicarbonate** is sodium hydrogen carbonate, $NaHCO_3$, and it has the food code E500.

Some supermarkets also have their own brands of baking powder and there are other branded versions.

### 4.   YEAST EXTRACT

*The Chemistry*

In the UK, yeast extract invariably means Marmite®, which is the vegetarian equivalent of concentrated beef extract (sold as Bovril® or Oxo® cubes). What the makers of Marmite® have tried to do is produce a blend of ingredients that replicates the intense umami flavour of a meat extract, and to which they have added extra vitamins. It is good source of B vitamins, but it is not to everyone's taste.

### Marmite®

*Ingredients (as listed)*

| | |
|---|---|
| Yeast extract | Spice extract (contains celery) |
| Salt | Riboflavin |
| Vegetable extract | Folic acid |
| Niacin | Vitamin $B_{12}$ |
| Thiamin | |

**Yeast extract** is made from brewery yeast, heated so its enzymes break down into single amino acids, such as glutamic acid. These are extracted with water. The water is then removed to yield a paste with a rich umami flavour.

© Shutterstock

**Salt** is sodium chloride.

**Vegetable extract** is usually made from soya, which is rich in protein. It is extracted using hydrochloric acid, followed by neutralising, filtering, decolourising, and concentrating. It has a bouillon-like odour.

**Niacin,** also known as nicotinic acid, and used to be called vitamin $B_3$. A lack of this vitamin causes pellagra, a deficiency disease whose main symptoms are dermatitis and chronic diarrhoea.

**Thiamin,** also known as thiamine hydrochloride, is vitamin $B_1$. This vitamin is needed to process carbohydrates to release their energy and it is involved in the nervous system.

**Spice extract (contains celery).** The composition of this is not specified. Because some people are allergic to celery, this has to be indicated according to EU law.

**Riboflavin** is vitamin $B_2$. This vitamin is important for keeping the skin and eyes in good condition, as well as being essential in foetal development.

**Folic acid,** also known as folate, and used to called vitamin $B_9$. It is vital to have a good supply of this vitamin before conception as it vital for brain development in the foetus. It also plays a role in several other bodily functions such the synthesis of DNA and the repair of RNA.

**Vitamin $B_{12}$** is also known as cobalamin. This is involved in the making of red blood cells and a lack of this can cause anaemia.

A similar product is Vegemite®, which also contains malt extract, potassium chloride, and the colouring agent ammonia caramel.

## 5.  MAYONNAISE

*The Chemistry*

Mayonnaise is a tasty addition to many foods and in particular to salads. It also can be used as a dip. Beware! It can provide a lot of calories. A typical mayonnaise will supply you with around 700 calories per 100 g, or 210 calories per typical portion of 30 grams.

Mayonnaise can be made at home from vegetable oil, egg yolk, and vinegar, plus small amounts of salt, sugar, or mustard to taste. These can all be blended together to form an emulsion, although this might separate into oil and water layers unless care is taken. It is much easier to buy a ready-made brand of mayonnaise, and then you can be certain it will remain as an emulsion until needed. It is also possible to buy a light mayonnaise version in which some of the oil has been replaced by starch, and in this way the calories it provides can be halved. This is the one described below. (There is also Hellmann's® Lighter-than-Light mayonnaise, which has only 20 calories per 30 grams.)

### Hellmann's® Light Mayonnaise

This product is flavoured with lemon juice, mustard, and paprika. The vegetable oil is mainly rapeseed oil.

*Ingredients*

| | |
|---|---|
| Water | Lemon juice |
| Vegetable oil (28%) | Mustard flavouring |
| Modified maize starch | Preservative (potassium sorbate) |
| Pasteurised egg & egg yolk (4.2%) | Stabilisers (guar gum, xanthan gum) |
| Spirit vinegar | Mustard |
| Salt | Antioxidant (calcium disodium EDTA) |
| Sugar | Paprika extract |
| Cream | |

**Vegetable oil** is not specified as such but is likely to be one of the edible oils such as sunflower oil.

**Modified maize starch** is also known as modified cornflour. This is starch that has been treated so that it will act as a thickening agent.

**Pasteurised egg & egg yolk (4.2%)** is produced by heating at 72 °C for 15 seconds and was devised by Pasteur as a method which ensured the killing all bacteria in milk.

**Spirit vinegar** is a solution of acetic acid produced from fermented cane sugar or molasses and then distilled to purify it.

**Preservative (potassium sorbate)** suppresses any microorganisms that might contaminate the mayonnaise, and especially ones that are introduced once the jar is opened, and why it should always be stored in a fridge.

**Stabilisers**. These are **guar gum** and **xanthan gum** that keep the ingredients in the form of an emulsion and prevent them separating into oil and water layers.

**Antioxidant (calcium disodium EDTA)** is there to deactivate metal ions and so prevent them reacting chemically with other ingredients.

Heinz also produces a light mayonnaise, and there are several own-brand supermarket mayonnaises as well.

## 6. STOCK

*The Chemistry*

You can make stock by simmering skin, bones, and leftover meat for a long time. It consists mainly of protein, which comes from the breakdown of collagen, the connective tissue. Other ingredients can be added, such as onion to give the stock flavour, and the process can be speeded up by using a pressure cooker.

Because making stock is time consuming, many prefer to buy a commercial version and use that for making gravies, pasta dishes, and soups.

### Knorr® Chicken Stock Pot

This is a popular type of instant stock and its packaging states that it contains no artificial flavour enhancers and specifically says there is no *added* MSG (although it does contain some of this), no artificial preservatives, no artificial colours, and is gluten free.

*Ingredients (as given on the pack)*

| | |
|---|---|
| Water | Leek |
| Glucose syrup | Parsley |
| Salt | Gelling agents (xanthan gum, |
| Sugar | locust bean gum) |
| Flavourings | Garlic |
| Lower sodium natural mineral salt* | Chicken powder (0.2%) |
| Yeast extract | Colours (plain caramel, mixed |
| Chicken fat (2.1%) | carotenes) |
| Carrots | Maltodextrin |
| Vegetable fat | Carrot juice concentrate |

*Contains naturally occurring potassium.

**Glucose syrup**, **salt**, and **sugar** together constitute most of the ingredients of this product.

**Flavourings** refers to all the various ingredients that add flavour.

**Lower sodium natural mineral salt** contains some salt from the Dead Sea, which is richer in magnesium chloride than in sodium chloride, and it also contains potassium chloride.

**Yeast extract** is made from brewery yeast, heated so its enzymes break down into single amino acids, such as glutamic acid. These are extracted with water.

**Chicken fat (2.1%)** is what it says.

The vegetables, herbs, and spices in the stock are **carrots, leek, parsley,** and **garlic**.

**Vegetable fat** is vegetable oil that is solid at room temperature and so classed as fat.

**Gelling agents** are there to ensure that the stock remains jelly-like and this is achieved by means of **xanthan gum** and **locust bean gum**.

**Chicken powder (0.2%)** is also a product of Knorr and appears to be added here. (Chicken powder itself consists of salt, flavour enhancers, namely monosodium glutamate (food code E621), disodium inosinate (E631), and disodium guanylate (E627), plus cornflour, sugar, chicken meat and fat (8%), egg, yeast, spices, and two colours, sulfite ammonia caramel (E150d) and tartrazine (E102).)

The **colouring agents** are **plain caramel** and **mixed carotenes**. The former is made by boiling sugar, the latter is either made by industry or extracted from fungus.

**Maltodextrin** is an easily digestible carbohydrate made from wheat starch.

**Carrot juice concentrate** provides colour.

There are other brands of instant stock available, such as Kallo®.

## 7.   GRAVY GRANULES

*The Chemistry*

Stirring cornflour into the juices from a roast joint will produce excellent, thickened gravy. (The cornflour needs to be mixed with a little water first.) Cornflour is mainly starch obtained from maize (called corn in the USA) and it has the benefit of dissolving to give a translucent mixture whereas ordinary flour would make it opaque. As starch is heated in the gravy its polymer chains, which are made up of glucose units, unravel and link up with other chains, thereby thickening the liquid to give the consistency that the cook is aiming for.

Alternatively you can use gravy granules to produce an instant gravy.

### Bisto® Best Rich Roast Beef Gravy

This is made simply by putting four or more heaped teaspoons, depending on taste, of the product into 280 ml (half a pint) of boiling water while stirring.

© Shutterstock

*Ingredients (as given on the pack)*

| | |
|---|---|
| Potato starch | Flavour enhancers (E621, E635) |
| Maltodextrin | Beef extract |
| Salt | Emulsifier (E322) |
| Flavourings | Spice & herb extracts |
| Vegetable oil | Onion extract |
| Colour (E150c) | |

**Potato starch** is just what it says.

**Maltodextrin** is made from wheat starch and gives a smoother feel in the mouth.

**Salt** is sodium chloride.

**Flavourings** is fourth on the list because it totals all the constituent flavours.

**Vegetable oil** is unspecified but is probably rapeseed or corn oil.

**Colour (E150c)** is ammonia caramel and is made by heating various sugars with ammonium hydroxide to give a dark brown colour.

**Flavour enhancers (E621, E635).** The first of these is monosodium glutamate (food code E621), which enhances the umami taste. The second is a mixture of disodium inosinate and disodium guanylate, which also enhance the savoury flavour.

**Beef extract** is beef stock that has been reduced to a thick paste.

**Emulsifier (E322)** is lecithin, which is a natural component of animal tissue. It can also be extracted from soybeans and rapeseeds. It is there to ensure the gravy has a consistent texture and is not lumpy.

**Spice & herb extracts** are not specified.

**Onion extract** is obtained by crushing onions and extracting the flavour molecules with water.

There are other gravy granules, for example, chicken, lamb, onion, vegetable, and turkey gravy. Other brands such as Schwartz® instant gravy granules are also available.

## 8.  WASHING-UP

*The Chemistry*

Washing-up aims to remove all kinds of food residues and some of this may be congealed fat. Therefore we need the help of a surfactant: a molecule with one end that is attracted to fat, the other attracted to water and which has the ability to dislodge the fat and solubilise it in the water. In fact it is not made soluble as such but forms microcelles that remain in the water.

Other components in a washing-up liquid may be there to prevent scum forming, to generate foam – which we can use as an indicator of how much washing-up liquid remains active – and perhaps to provide a pleasant aroma.

### Fairy® Liquid Platinum

The most popular washing-up product is Fairy® Liquid, of which their premium brand is Fairy® Platinum. The bottle gives the following information: 15–30% anionic surfactants, 5–15% non-ionic surfactants, methylisothiazolinone, phenoxyethanol, perfumes, butylphenyl methylpropional, hexyl cinnamal, and limonene. Their website gives a complete list of ingredients:

*Contents*

| | |
|---|---|
| Aqua | Parfum |
| Sodium laureth sulfate | Phenoxyethanol |
| Lauramine oxide | Sodium hydroxide |
| C9–11 pareth-n | Butylphenyl methylpropional |
| SD alcohol 40 | Hexyl cinnamal |
| Sodium chloride | Limonene |
| Tetrasodium dicarboxymethyl   glutamate | Methylisothiazolinone |
| | Colourant |
| PPG-34 | Colourant |

**Aqua** is water and solvent.

**Sodium laureth sulfate** is an anionic surfactant made from coconut oil. It is gentle and lathers well.

**Lauramine oxide** and **C9–11 pareth-n** are non-ionic surfactants.

**SD alcohol 40** controls the viscosity and clarity of the liquid. This is ethanol that has been made undrinkable by the addition of Bitrex®.

**Sodium chloride** acts as a thickener.

**Tetrasodium dicarboxymethyl glutamate** is also known as tetrasodium glutamate diacetate. It is a chelating agent and there to stop calcium in hard water from forming a scum.

**PPG-34** is polypropylene glycol and it is a skin conditioner to protect the hands.

**Parfum.** Perfume raw material is a complex mixture, items of which have to be identified in a label because some people may be sensitive to them. The ones so identified here are **butylphenyl methylpropional**, which has an outdoor fresh smell; **hexyl cinnamal**, which smells of camomile; and **limonene**, which smells of oranges.

**Phenoxyethanol** is there to prevent microbes breeding in the washing-up liquid.

**Sodium hydroxide** raises the pH above 7 and thereby helps dissolve grease.

**Methylisothiazolinone** is an antimicrobial agent.

Two colourants are used in this product but neither is identified.

There are other versions of Fairy® Liquid, and other brands of washing-up liquids such as Persil®, supermarket own-brands, and Ecover®.

## Dishwasher Tablets

### The Chemistry

Dishwasher tablets have to remove all kinds of food residues from all kinds of surfaces. Food residues can be carbohydrates, oils, fats, proteins, and stains, and surfaces can be crockery, glass, plastic, and metal. The tablets need to contain a surfactant, a water softener, enzymes to digest food residues, a bleach to remove stains, and the final rinse must not leave drops of water that dry out causing marks, especially noticeable on glassware. Ideally all these should be contained within a single tablet and chemists have been able to devise such tablets, as the following product shows.

© Shutterstock

**Finish® Quantum®**

In this product there are three compartments (white, red, and blue) and this is necessary because the various chemicals that they contain are either not compatible with one another or need to be released at different rates. In the white compartment, which is released first, there is a fine powder that is mainly a mixture of water softeners and bleaching agents; in the red central compartment are the enzymes and they are released second; and in the third, blue, compartment, is a gel with a non-ionic surfactant, and this dissolves last of all. The enzymes need to be kept separate from the bleaching agents, at least until they are in the wash water when both kinds of agents get to work. Finally they are joined by other chemicals from the blue compartment and these are there to ensure that water drains from surfaces without leaving drops.

*Contents*

Pentasodium triphosphate
Sodium carbonate peroxide
Polyvinylalcohol
Fatty alcohol alkoxylate
2-Propenoic acid, homopolymer,
  sodium salt, sulfonated
Aqua

Tetrasodium etidronate
PEG-800
2-Propenoic acid, homopolymer,
  sodium salt
Sodium carbonate
TAED
Sorbitol

Propylene glycol
Sodium sulfate
Trimethylolpropane
Cellulose gum
Sucrose
*Oryza sativa* starch
Sodium chloride
Dimethicone
Methyl-1H-benzotriazole
Protease
PEG 130–PEG 150
Parfum
Manganese(II) oxalate

Soap
Stearamide
Colourant
Citric acid
Petroleum distillates
C12–13 pareth-6
Amylase
Steareth-21
Titanium dioxide
Cellulose
Kaolin
Diallyldimethylammonium chloride

## Water Softeners

**Pentasodium triphosphate**, and **tetrasodium etidronate** act as chelating agents to prevent metal ions from interfering with the cleaning process.

**2-Propenic acid, homopolymer, sodium salt sulfonated** and **2-propenoic acid, homo-polymer, sodium salt** are polymers acting as water softeners and to keep food traces in suspension in the wash water.

**Sodium carbonate** makes the wash water slightly alkaline and helps to remove grease.

## Bleach

**Sodium carbonate peroxide** is a bleaching agent able to remove tea, coffee, ketchup, sauce, fruit, and wine stains. To do its job, the peroxide needs **TAED** (tetraacetylethylenediamine) with which it reacts to form sodium peracetate, which is effectively the real bleaching agent. **Manganese(II) oxalate** acts as a catalyst for this reaction.

## Packaging

**Polyvinylalcohol** is a water-soluble polymer from which the tablet and its components are made. It is very soluble in water and dissolves in a pre-planned sequence.

## Surfactants

**Fatty alcohol alkoxylate** is a low foam, non-ionic surfactant that removes grease yet does not produce a lot of unnecessary foam. The same is true of **C12–13 pareth-6**.

**Dimethicone** is silicone, which acts as an anti-foaming agent.

*Corrosion Inhibitor*

**Methyl-1H-benzotriazole** is a corrosion inhibitor. It forms a protective film one molecule thick on the surface of metals like copper and silver and thereby protects them.

*Enzymes*

**Protease** is an enzyme that digests protein.
**Amylase** is an enzyme that removes carbohydrates.

*Other Ingredients*

The various other components of the contents list are auxiliary items that come with some of the raw materials, for example as protective coatings, or are added as solvents as is **propylene glycol**, or are structural components such as **PEG-130, PEG-150,** and PEG-800.
**Sorbitol** acts as a humectant.
**Sodium sulfate** is a filler.
**Trimethylolpropane** is a component of the polymers and added to cross-link them.
**Cellulose gum** is a thickening agent.
**Sucrose** is a thickening agent.
*Oryza sativa* **starch** is a viscosity modifier.
**Sodium chloride** acts as a thickener.
**Soap** is a detergent.
**Stearamide** has fabric softener properties.
**Citric acid** makes the liquid acidic.
**Petroleum distillate** is a hydrocarbon solvent.
**Stearth-21** is an emulsifying agent.
**Titanium dioxide** is a white pigment.
**Cellulose** is part of the packaging
**Kaolin** is a soft mineral and is a filler.
**Diallyldimethyammonium chloride** is an antimicrobial agent.
**Parfum** is not specified nor is **colourant** but these are added for aesthetic reasons.
**Aqua** (water) is present in the gel.

Other dishwasher tablets include supermarket own-brands and Ecozone®. There is also a Finish® dishwasher liquid, but this requires the dishwasher to be primed with a separate rinse aid.

## 9. OVEN CLEANER

*The Chemistry*

There are two kinds of cleaner able to remove the burnt-on grime that collects inside an oven. They are a powerful surfactant and a strong alkali. Both take time to work because they have to penetrate a baked-on surface layer.

### Mr Muscle® Oven Cleaner

This comes in an aerosol pack so that you can spray the cleaner over the grime in the oven (which must not be hot), then leave for up to two hours, before wiping it away to reveal a clean oven.

*Contents (as given on the pack)*

| | |
|---|---|
| Aqua | Sodium alkyl C9–11 sulfate |
| Sodium hydroxide | Parfum |
| Butane | Limonene |
| Propane | Sodium lauryl sulfate |
| Acrylic copolymer | Methylchloroisothiazolinone |
| Isobutane | Methylisothiazolinon |

**Aqua** is water and solvent.

**Sodium hydroxide** is a powerful alkali able to react with grease, thereby making it soluble. It also attacks burnt-on deposits and solubilises them.

**Butane**, **propane**, and **isobutane** are the propellants. These are under pressure in the canister and they force out the content through a nozzle when the pressure is released.

**Acrylic copolymer** acts as a thickener for the other ingredients so they don't run off the oven surfaces and it helps to break down the grease.

**Sodium alkyl C9–11 sulfate** and **sodium lauryl sulfate** are anionic surfactants.

**Parfum.** Perfume raw material is a complex mixture, some items of which have to be identified in a label because some people may be sensitive to them. The one so identified here is **limonene,** which smells of oranges.

**Methylchloroisothiazolinone** and **methylisothiazolinone** act as antimicrobials.

Another example of oven cleaner is Oven Pride™.

## 10.   MATCHES

*The Chemistry*

When most cookers were fuelled by gas, matches were a popular way of lighting them. Now they are rarely needed in the kitchen, although they may be needed to light candles.

There are strike-anywhere matches in which the chemical that initiates the flame is part of the match head so that friction alone will start it burning. That chemical is phosphorus sulfide. However, most matches are safety matches that can only be ignited by striking the match head against the side of the box that contains the initiator, which is red phosphorus.

The first stage of burning involves a strong oxidising agent, namely potassium chlorate, reacting with sulfur. The flame then transfers to wax that is below the match head, and then to the wooden match itself. This wood has also been treated with ammonium phosphate, which prevents afterglow.

### The Original Cook's® Matches

These are safety matches.

*Components*

| *Match heads* | *Striking surface* |
|---|---|
| Ammonium dihydrogen phosphate | Red phosphorus |
| Paraffin wax | Powdered glass |
| Potassium chlorate | Glue |
| Sulfur | |
| Powdered glass | |
| Pigment | |
| Glue | |

**Ammonium dihydrogen phosphate** prevents afterglow, so that when the flame goes out it does not leave behind a smouldering match capable of igniting something.

**Paraffin wax** is used to coat the tip of the match stick. It acts as the first fuel to be ignited by the flame. This then ignites the wood of the matchstick.

**Potassium chlorate** starts a chemical reaction with the sulphur when the match head is struck and this produces a flame.

**Sulfur** reacts with the potassium chlorate violently enough to cause combustion.

**Powdered glass** increases the friction between the match head and the matchbox, and this generates a localised temperature of around 200 °C, which is needed to cause the chemical reactions to begin.

**Red phosphorus** is the stable form of this element but the friction of striking the match converts it to white phosphorus, which is spontaneously flammable and so ignition begins.

**Glue** is needed to fix hold together the ingredients of the match head, and those of the striking surface, and attach the former to the matchstick and the latter to the matchbox.

There are other brands of match, of which Swan Vestas® are a strike-anywhere type. However, many of the traditional brands of matches are no longer available, such as Bryant & May, Captain Webb, and England's Glory.

## 11.  SMOKE ALARM

*The Chemistry*

Smoke alarms are an essential part of the home and can be located throughout the house. The type described here relies on radioactivity to do the job we expect of it.

A smoke alarm contains less than a microgram of americium oxide $(AmO_2)$, a man-made radioactive element. Even this tiny amount releases around 30 000 alpha particles a second whilst emitting almost no gamma radiation. (A sheet of paper will stop an alpha particle but gamma radiation is very penetrating.)

When an alpha particle collides with oxygen or nitrogen it grabs negative electrons from it and in so doing forms positive ions. They respond to a voltage difference across a gap and move towards the negative electrode, in effect causing an electric current to flow. When smoke enters the gap it absorbs the ions and the current drops. This is detected and the alarm is triggered. (If you are confident that the situation is safe you can waft the smoke out of the detector to restore the device to its normal state.)

Ideally a smoke alarm should draw its current from the domestic supply but with battery back-up, and this should be checked occasionally. Those smoke alarms that are purely battery operated should be checked much more frequently. This can be done by pressing a button that will trigger the alarm to sound.

© Shutterstock

## Components

| | |
|---|---|
| Americium-241 | Battery |
| Silver | Alarm |
| Gold | Polymer casing |
| Printed circuit board | |

**Americium-241** is a radioactive man-made element.

**Silver** is used as the main backing support for the radioactive discs.

**Gold** serves two roles. The active layer is a foil made of gold and americium oxide, and coated both sides with a layer of gold, and then attached to the silver backing. Finally the disc is given a layer of gold plate. Very little gold is actually needed and the amount of gold in a smoke alarm is worth less than 5 pence.

The **printed circuit board (PCB)** contains the electrical components like diodes, capacitors, and resistors that are needed to measure and compare the electric current in the sensor with that in a sealed reference chamber. If there is a difference then it triggers off the alarm. The PCB also monitors the condition of the battery and is programmed to issue a warning in the form of a short signal at minute-long intervals to indicate when the battery needs replacing.

The **battery** has to be the 9 volt long-life type.

The **alarm** has to be capable of emitting a sound that can be heard several metres away, and is typically around 90 decibels in volume – equivalent to the noise a motorbike makes. It has to be capable of waking someone in another room who is in a deep sleep.

The **polymer casing** is made of PVC, which is a tough, long-lasting, fire-resistant plastic.

## 12.   KETTLE AND IRON DESCALER

*The Chemistry*

Most homes in the UK are in areas where the domestic supply of water is classed as hard, which means it contains dissolved calcium and magnesium ions in the form of calcium hydrogen carbonate and magnesium hydrogen carbonate. When such water is boiled in a kettle, or heated in an iron, it forms calcium carbonate and magnesium carbonate. Both of these are insoluble and they precipitate out – that's what limescale is. To re-dissolve these deposits we need to attack them with an acid, whereupon the carbonate is converted to carbon dioxide and the calcium and magnesium are converted to a soluble salt.

While some liquid acids, like hydrochloric acid, would do this very well they might also attack the metal itself. In any case, it may not be advisable to put such an acid into the hands of the general public, because, in more concentrated forms, it has a corrosive effect on human tissue. The alternative is a weak acid like vinegar (acetic acid), or better still is a solid acid. This is added to the appliance with enough water to dissolve it, and the solution will then attack the limescale, although it might take some time if the limescale is a rather thick layer. When it has achieved the desired result, the solution can safely be poured away down the sink, the appliance rinsed, and then reused.

### Ecozone® Kettle and Iron Descaler

*Contents*

Citric acid (100%).

**Citric acid** is a powder that is to be dissolved in water and then used. It will attack the limescale by the chemical process:

$$\text{acid} + \text{carbonate} \rightarrow \text{salt} + CO_2 \uparrow$$

Waitrose own-brand descaler comes as a solution of citric acid, which can be used as directed and left overnight. There are descalers based on other acids, for example, Oust® uses lactic acid and this is also a solid.

CHAPTER 9

# The Dining Room/Food and Drink

© Shutterstock

Chemistry at Home: Exploring the Ingredients in Everyday Products
By John Emsley
© John Emsley, 2015
Published by the Royal Society of Chemistry, www.rsc.org

In this chapter we will look at the ingredients in foods and drinks that contain additives, and that applies particularly to products that promise to have fewer calories. Low-sugar alternatives, which substitute sweeteners for sugar, will be a key feature of this chapter – although even these are not without critics. Most of this criticism came as a result of research in the 1970s that suggested cyclamate and saccharin caused bladder cancer, which led to them being banned in the US and elsewhere. However, later research found no support for these research claims, and the sweeteners were reinstated and are now approved for use by various government and international agencies. Other sweeteners have suffered adverse comments, and these can be found on various websites, but again further investigation has proved such claims to be alarmist and unjustified.

Of course the food you eat may contain additives, such as colourants and preservatives. In the last century, these came in for a great deal of criticism despite the fact that they were given E numbers, which meant they are regarded as safe in every country of the EU.

E-labelled additives come in several categories, of which the main ones are colours, preservatives, antioxidants, thickeners/emulsifiers, acidity/pH regulators, anti-caking agents, flavour enhancers, and sweeteners. Some are undergoing a major change within the food industry, namely food colours, preservatives, and flavours. There is now a desire to remove synthetic dyes from foods and replace them with natural colourants, so that red, orange, and yellow dyes are being replaced by colourants that Nature produces, namely carmine (E120), carotene (E160), and turmeric yellow (E100). (Carmine may not be to everyone's taste as it comes from the cochineal insect.) Blue is more of a problem, although a new blue colourant, phycocyanobilin, can be extracted from the cyanobacteria *Arthrospira platensis* and may one day be used.

The acceptable antioxidants that can be used are BHT (short for butylated hydroxytoluene, E321), BHA (butylated hydroxyanisole, E320) and TBHQ (*tert*-butyhydroquinone, E319). These are powerful antioxidants that prevent oil from going rancid. Antimicrobials are added to ensure that moulds and bacteria don't enjoy your food before you do. Potassium sorbate (E202) and sodium benzoate (E211) are generally regarded as safe but even they may be replaced by naturally derived chemicals one day, and there are rosemary extracts, rosmarinic acid and carnosic acid, which may offer the same kind of protection.

Humans can distinguish five types of taste: sweet, sour, bitter, salty, and savoury (often referred to as umami). The molecules that produce these sensations on the tongue come from three sources: there are natural ones, nature-identical ones, and artificial ones. Several of these will feature in the products we will be looking at.

One flavour, vanilla, poses a particular problem. Worldwide vanilla production from plants amounts to around 2000 tonnes a year but not all is used as a flavouring agent. The major source of vanilla flavour is the molecule vanillin and most of this (10 000 tonnes a year) is produced by the chemical industry and is a hundred times cheaper than natural vanillin. Of course it is exactly the same molecule as the natural vanilla, although for some people this simply makes it a 'chemical' hence the trend is to imply that a vanilla-flavoured product contains at least a modicum of the natural extract.

I have also included some dietary supplements in this chapter, which many people now take regularly in the belief that their normal diet is lacking in some way. This might well be true for some who persist in always eating the same foods day after day. For most people the advice to eat five pieces of fruit or five vegetable portions a day (and not counting potatoes) should ensure the body gets enough of the micronutrient vitamins and minerals that these provide.

The categories in this chapter are as follows, and they deal mainly with food over which concerns have been expressed:

1. Butter and margarine (Lurpak® Lightest; Flora™; Flora™ Pro-Activ® Light)
2. Sweeteners (Sweetex®; Canderel®; Sweet'n Low® Blend; Splenda®; Truvia®)
3. Refreshing drinks (Diet Coke®; Irn-Bru®; Irn-Bru® Light; Drench® Cranberry and Raspberry Drink)
4. Energy drinks (Relentless®, Lucozade® Energy Original, Quick Energy™)
5. Hot drinks (Cadbury Highlights®, Coffee-mate®)

6. Desserts (Shape® Strawberry Yoghurt; Hartley's® Sugar Free Jelly; Carte D'Or® Ice Cream; Carte D'Or® Light)
7. Vitamins (Berocca® Vitamins and Minerals; Boots® Selenium with Vitamins A, C, and E; Boots® Cod Liver Oil and Calcium)

## 1. SPREADS

### Butter

*The Chemistry*

The traditional spread for bread is butter. This is made from cream and consists of 85% fat, 15% water, and traces of various flavour molecules. The fat is 67% saturated, 28% monounsaturated (mainly oleic fatty acid), and 5% polyunsaturated (mainly linoleic fatty acid). It is the saturated content that causes some people to prefer margarine or a blend of butter and vegetable oil, or with water added, which reduces its calorie-content dramatically.

Normal butter provides 75 calories per 10 gram portion (which is the size of those small packs provided in restaurants), Lurpak Spreadable is about the same at 73 calories. It is 69% butter, 25% vegetable oil to help it spread more easily when cold, plus a little water, salt, and lactic culture. Of course, if you are watching your weight and still want to eat bread with butter there is a Lurpak version that has even fewer calories – only 38 in a 10 gram portion. This is the one discussed here.

© Shutterstock

## Lurpak® Lightest

*Ingredients*

| | |
|---|---|
| Water | Milk protein |
| Butter (27%) | Salt (1%) |
| Vegetable oil (19%) | Preservative (potassium sorbate) |
| Lactic culture | |

**Water** replaces more than half of the high-calorie fat.

**Butter (27%)** is just what it says.

**Vegetable oil (19%)** ensures that the mixture has a lower softening temperature so it can be spread straight from the fridge. The oil is rapeseed oil.

**Lactic culture** is used to speed up the process of forming butter once the milk has been pasteurised. (Lactic acid will form naturally in unpasteurised milk.)

**Milk protein** is mainly casein and it acts to emulsify the other ingredients.

**Salt (1%)** is sodium chloride.

**Preservative (potassium sorbate)** (food code E202) is a highly effective fungicide that stops moulds and yeasts.

## Margarine

*The Chemistry*

Butter maybe natural but it is high in saturated fats and was said to increase the level of 'bad' cholesterol in the blood, which increases the likelihood of a stroke or heart attack. The research which supported these claims has since been criticised as being compromised by confounding variables, and eating butter is now said not to be a factor in these conditions.

The alternative spread is margarine and the makers of this can claim it too is natural, being made from plant oils such as olive oil, and so is a mainly unsaturated fat having both mono-unsaturated (oleic), and polyunsaturated (linoleic).

However, there are things missing from margarine and they are the fat-soluble vitamins, namely A, D and E – so these need to be added, along with folic acid. Margarine can be given a butter-like flavour by using buttermilk, which is the watery layer left behind when cream is churned into butter.

Plants produce oils but these are not convenient for spreading on bread and crackers. However, oils and fats are chemically the same and the former can be converted to the latter by reaction with hydrogen gas, which thereby raises the melting point so what was an oil becomes a fat. This process removes some of their double bonds, converting CH groups along their fatty acid chains to $CH_2$ groups. Unfortunately this chemical process can also change the structure of some of the hydrocarbon chains, creating so-called 'trans' fats, which are deemed to be a health risk. Food chemists have been able to devise hydrogenation processes that don't yield these kind of fats in margarine. Here we look at one popular brand of margarine: the margarine itself, and a diet version with fewer calories.

### Flora®

This margarine includes buttermilk to provide a similar flavour to butter. It provides 53 calories per 10 gram portion. Flora® margarine also contains omega-3 and omega-6 oils, which have health benefits and come with the seed oils used to make the margarine.

*Ingredients*

Water
Vegetable oils (45%)
Buttermilk (10.5%)
Salt (1.4%)
Emulsifiers: mono- and di-glycerides
  of fatty acids, Sunflower lecithin
Natural flavouring (contains milk)
Preservative: potassium sorbate
Citric acid
Vitamin A
Vitamin D
Colour (carotene)

**Water** accounts for a significant proportion of this spread.

**Vegetable oils (45%)** can be one of various edible oils, such as sunflower, linseed, palm, or rapeseed oil. Sunflower oil accounts for 26%.

**Buttermilk (10.5%)** is the watery liquid left behind after cream has been churned into butter. It contains soluble proteins and lactose (also known as milk sugar).

**Salt (1.5%)** is sodium chloride.

**Emulsifiers: mono- and di-glycerides of fatty acids** (food codes E471 and E472) and **lecithin** (food code E322) are emulsifiers, added to ensure that all the ingredients blend to a smooth consistency.

**Natural flavouring** is a term that can cover all kinds of possible flavour components that have been derived from spices, fruits, vegetables, yeast, roasted meats, eggs, milk, *etc*. In the case of Flora they will probably have been derived from dairy products.

**Preservative: potassium sorbate** (food code E202) protects the margarine from becoming infected with moulds and yeasts.

**Citric acid** (food code E330) is produced naturally by citrus fruits but is more cheaply manufactured from starch. Here it acts as a preservative.

**Vitamin A** is an oil-soluble vitamin needed for growth, development, the eyes, and the immune system.

**Vitamin D** is an oil-soluble vitamin that is needed to help the body absorb mineral nutrients like calcium.

**Colour (carotene)** is the yellow colouring agent. It is extracted from carrots and oranges.

## Flora® Pro-Activ™ Light

This version of the above margarine offers something extra. It includes sterol esters, which prevent the body actually absorbing cholesterol. Some cholesterol is produced by the liver and excreted into the intestines in order to digest fats and oils, after which it is reabsorbed into the blood stream. Sterol esters block the reabsorption sites.

This spread provides only 32 calories per 10 gram portion as opposed to Flora® original, which provide 53 calories per 10 gram portion.

### *Ingredients*

Water
Vegetable oils (30%)
Plant sterol esters (12.5%)
Buttermilk
Modified waxy corn starch
Salt (1.0%)
Emulsifier: mono- and di-glycerides of
   fatty acids and sunflower lecithin

Preservative (potassium sorbate)
Citric acid
Flavourings
Colour: (beta-carotene)
Vitamins A and D

**Water** is the major component of this product and hence its low calorie content.

**Vegetable oils (30%)** can be one of various edible oils, such as sunflower oil, rapeseed oil, or olive oil.

**Plant sterol esters (12.5%)** (also known as phytosterol) is the ingredient behind the cholesterol-lowering claim explained above.

**Buttermilk** is the watery liquid left behind after cream has been churned into butter. It contains soluble proteins and lactose (also known as milk sugar).

**Modified waxy corn starch** is starch guaranteed free of gluten and it is used as a thickener and emulsifying agent. This is starch that has been treated with acid to break its chains into shorter lengths.

**Salt (1.0%)** is sodium chloride.

**Emulsifiers: mono- and di-glycerides of fatty acids** (food codes E471 and E472) **and sunflower lecithin** (food code E322) act as emulsifiers to ensure the water and vegetable oils remain consistently mixed.

**Preservative: potassium sorbate** (food code E202) is a preservative that protects the margarine from becoming infected with moulds and yeasts.

**Citric acid** (food code E330) acts as a preservative.

**Flavourings** are not specified.

**Colour: beta-carotene** is the yellow colouring agent. It is extracted from carrots and oranges.

**Vitamins A and D** are oil-soluble vitamins needed for growth, the eyes, and the immune system (vitamin A) and to help the body absorb mineral nutrients like calcium (vitamin D).

## 2.   SWEETENERS

*The Chemistry*

Sweetness is one of the five basic tastes. Indeed, for many, it is probably the most enjoyable taste of all and one that dominates our childhood preferences. In Nature it indicates a source of food rich in easily digestible carbohydrate, and so high in energy.

The trouble with sweetness is that it is associated with several foods that are also rich in calories, such as sugar, syrup, and honey, and the foods made from them such as sweets, biscuits, icing, and cakes. For these foods it is not easily replaced, but there are occasions when we can use an artificial sweetener instead, such as in tea, coffee, colas, and other drinks. There are several intense sweeteners, namely saccharin, cyclamate, aspartame, acesulfame, and sucralose, and a natural one called stevia. (This last one is extracted from the *Stevia rebaudiana* plant that is native to South America.) These molecules were discovered to trigger the sweetness receptors on the tongue and they offered a way of enjoying this sensation without adding excess calories to the diet.

Most people know artificial sweeteners by their trade names and these are the ones described below. Their tiny tablets are designed to provide the sweetness equivalent to a teaspoonful of sugar.

## Sweetex®

This is saccharin and is the oldest artificial sweetener, used for more than a century. This brand is the most popular version, but there are other saccharin-based sweeteners such as Hermesetas®.

*Ingredients*

Sodium saccharin
Silicon dioxide
Magnesium stearate

**Sodium saccharin** is the intense sweet-tasting chemical; its food code is E954.

**Silicon dioxide** is an anti-caking agent.

**Magnesium stearate** is necessary to ensure the ingredients do not stick to the machine that stamps out the tablets.

## Canderel®

This is aspartame, which is a combination of two amino acids in the form of a methyl ester. In this product it is blended with another artificial sweetener, acesulfame K, because each intensifies the sweetness of the other, so less of either is needed. Because a few people – one person in 15 000 – are sensitive to one of the amino acids (phenylalanine), this has to carry a warning.

Such individuals have a genetic disorder known as phenylketonuria and it means they cannot metabolise this amino acid in the normal way, so have to regulate their protein intake carefully to control its intake. In phenylketonuria the body converts phenylalanine to phenylketone and this substance can cause brain damage.

### Ingredients

| | |
|---|---|
| Aspartame | Binder: cross-linked CMC |
| Acesulfame K | Flavouring |
| Leucine | |

**Aspartame** and **acesulfame K** are intense, calorie-free, sweeteners that together intensify each other's effect, so less of each is needed.

**Leucine** (food code E641) is an essential amino acid from which protein is created. It is essential because it has to be part of our diet since we cannot synthesise it within the body, as we can with most amino acids. It is added here as a lubricant or binder.

**Binder: cross-linked CMC** (carboxymethyl cellulose or cellulose gum) acts to hold the other ingredients together, but easily breaks up in water to release the sweeteners.

**Flavouring** is not specified.

## Sweet'n Low® Blend

There are slightly different formulations of Sweet'n Low®, which come in the same small packaging. This is a typical one. When the artificial sweeteners acesulfame K and aspartame are used together then each enhances the sweetness of the other, so less is needed of either. Sweet'n Low® contains only 7 mg of each one.

It comes either as tablets or packets of the powdered form, the latter having the composition:

*Ingredients*

| | |
|---|---|
| Dextrose | Aspartame |
| Maltodextrin | Silicon dioxide |
| Acesulfame K | Flavouring |

**Dextrose** is another name for glucose, and here it is used as a filler.

**Maltodextrin** consists of short chains of linked glucose molecules (up to around 16 in the longer chains) and it is made from wheat starch.

**Aspartame** and **acesulfame K** are intense, calorie-free, sweeteners that together intensify each other's effect, so less of each is needed.

**Silicon dioxide** acts as a filler.

**Flavouring** is not specified.

## Splenda®

This artificial sweetener (sucralose) is based on the sugar molecule (sucrose) in which three chlorine atoms have replaced three hydroxy (OH) groups. This not only intensifies the sweetness a thousand times, it also makes the molecules non-digestible so it cannot release any calories. The chlorine atoms make it impossible for enzymes to break sucralose down and so it passes through the body undigested. It is stable to heat and so can be used in cooking and baking. The version described below is the tablet form for adding to hot drinks, and it can also be obtained in a granuated form for sprinking over other foods and for use in cooking and baking.

*Ingredients*

| | |
|---|---|
| Lactose (from milk) | Leucine |
| Sucralose 11% | Cross-linked cellulose gum |

**Lactose (from milk)** is also known as milk sugar and is added as a filler to bulk out the tablet.

**Sucralose 11%** is the artificial sweetener.

**Leucine** (food code E641) is an essential amino acid, added here as a lubricant or binder.

**Cross-linked cellulose gum** holds the tablet together but easily breaks up in water to release the sucralose sweetener.

## Truvia®

This contains stevia, which is a naturally occurring intense sweetener, around 250 times sweeter than sugar.

### Ingredients

Lactose (from milk)  
Steviol glycosides  
  (Stevia leaf extract)  
Flavouring  
Stabilisers: cross-linked cellulose gum  
Magnesium salts of fatty acids

**Lactose (from milk)** is also known as milk sugar and acts as a filler.

**Steviol glycosides (Stevia leaf extract)** is the intense sweetener. It takes a few seconds to register as sweet on the tongue.

**Flavouring** is not specified.

**Stabiliser: cross-linked cellulose gum** holds the tablet together, but easily breaks up in water and release the sweetener.

**Magnesium salts of fatty acids** are added to the mixture of ingredients as a lubricant so that when these are stamped into tablets they don't stick to the machinery.

### 3.   REFRESHING DRINKS

*The Chemistry*

The chemist, Joseph Priestley, made the first carbonated drink in 1767 using the $CO_2$ given off from the vats in which a local beer was being brewed in Leeds where Priestley lived. Such drinks have been popular ever since. Putting the gas under pressure greatly increases the amount of $CO_2$ that dissolves.

There are numerous carbonated drinks, most of which come in two varieties – with sugar as the sweetener, or with artificial sweeteners.

### Coca-Cola®

This popular refreshing drink is best drunk cold. It was concocted in 1887 by an Atlanta pharmacist, Dr John Pemberton, and he named it Coca-Cola® after the coca plant, which was a source of cocaine, and the cola nut, which provided caffeine. The inclusion of these natural ingredients ended a century ago and caffeine is now added in its pure form.

© Shutterstock

Although the flavourings are not listed they were revealed by Mark Prendergast in 1993 who claimed to have found a list of them in the 1887 notebooks of Dr John Pemberton, the man who invented Coca-Cola®. The main flavours were lime juice and vanilla, with orange, lemon, nutmeg, cassia, coriander and neroli being minor ones. However, the Coca-Cola® company now says this list does not reflect the composition of the modern product which is still a secret.

## Diet Coke®

This version of Coca-Cola® contains no sugar, and now appears to be the drink of choice.

*Ingredients*

Carbonated water
Colour (Caramel E150d)
Sweeteners (Aspartame,
  Acesulfame K)

Natural flavouring including caffeine
Phosphoric acid
Citric acid

**Carbonated water** is water in which $CO_2$ has been dissolved.

**Colour (Caramel E150d)** provides the colour of this drink.

**Sweeteners (Aspartame, Acesulfame K)** are intense, calorie-free, sweeteners that together intensify each other's effect, so less of each is needed.

**Natural Flavourings including caffeine.** Caffeine has three effects: it stimulates the brain by increasing the level of dopamine; it relaxes the airways so making breathing easier; and it releases energy from stores within the body. The amount of caffeine in a can of cola is 40 mg. This is the same as the caffeine in a cup of tea, whereas a mug of fresh coffee has around 80 mg, although a mug of instant coffee has about 60 mg.

**Phosphoric acid** has no taste so does not interfere with the other flavours while making the drink more acidic.

**Citric acid** has a pleasing fruity tang.

## Irn-Bru®

Like Coca-Cola®, the ingredients that flavour this orange-coloured drink are surrounded in mystery. They are supposed to be known to only three people, who are sworn to secrecy. Irn-Bru® first appeared in 1901 and in the last century it was advertised as being 'made in Scotland from girders.'

*Ingredients*

Carbonated water
Sugar
Citric acid
Flavourings (including caffeine
  and quinine)

Preservative (E211)
Colours (sunset yellow and
  ponceau 4R)
Ammonium ferric citrate (0.002%)

**Carbonated water** is $CO_2$ dissolved in water under pressure.

**Sugar** is sucrose. Here – in a 330 ml can – there are 35 g, which provide 140 calories.

**Citric acid** is the acid of citrus fruits, with a pleasing fruity tang.

**Flavourings** are not listed but are 32 in number.

**Caffeine** has three effects: it stimulates the brain by increasing the level of dopamine; it relaxes the airways so making breathing easier; and it releases energy from stores within the body.

**Quinine** is a natural drug that reduces fevers and relaxes muscles, and it provides the drink with its bitter note.

**Preservative** (food code E211) is sodium benzoate, which is powerful antimicrobial agent, especially under acid conditions.

**Colours (sunset yellow and ponceau 4R)** are approved colourants. Sunset yellow has food code E110; ponceau 4R has food code E124 and is a deep red colour. Both are synthetic dyes.

**Ammonium ferric citrate (0.002%)** includes iron (as $Fe^{3+}$) and gives the drink its name. A 330 ml can contains about 0.1 mg, of which 0.015 mg is iron itself.

## Irn-Bru® Light

This sugar-free version contains only 2 calories per 330 ml, as opposed to 140 calories in the sugar-sweetened standard drink. This is made possible by replacing sugar with acesulfame K and aspartame. In all other respects the ingredients are the same as in the standard drink:

*Ingredients*

| | |
|---|---|
| Carbonated water | Sweeteners (acesulfame K, aspartame) |
| Citric acid | Preservative (E211) |
| Flavourings (including caffeine and quinine) | Colours (sunset yellow and ponceau 4R) |
| | Ammonium ferric citrate (0.002%) |

## Drench® Cranberry and Raspberry Drink

A 440 ml bottle of this drink contains a relatively high level of sugar, and consequently provides 145 calories. However, there is a diet form of this drink that has only 10 calories.

*Ingredients*

| | |
|---|---|
| Spring water | Citric acid |
| Grape 6% | Natural flavourings |
| Cranberry 1% | Anthocycanins |
| Raspberry 1% | Potassium sorbate |
| Sugar | Dimethyl dicarbonate |
| Stevia | |

**Spring water.** This originally was from an underground water source (spring or borehole) at Norwich, although it now appears to come from Huddersfield.

Fruit juices are **grape** juice 6%, **cranberry** juice 1%, and **raspberry** juice 1%, all of which are natural. The chief aroma molecule present in grapes is methyl anthranilate, those present in cranberries are ethyl-2-methylbutyrate, ethyl-3-methylbutryate, and *trans*-2-hexenal, while those in raspberries are 1-(*p*-hydroxyphenyl)-3-butanone, *cis*-3-hexene-1-ol, damascenone, $\alpha$-ionone, and $\beta$-ionone.

**Sugar** provides sweetness and calories.

**Stevia** is the intense natural sweetener. It takes a few seconds to register as sweet on the tongue.

**Citric acid** gives the drink added tanginess and acts as a preservative.

**Natural flavourings**: these are unspecified.

**Anthocycanins** act as powerful antioxidants. They are present in fruits such as red grapes, blueberries, and raspberries, which is why these fruits are thought to be particularly beneficial. Anthocyanins are coloured compounds (E163) that give many flowers their characteristic red, blue, or purple colours.

**Potassium sorbate** is a preservative (E202) and is a highly effective fungicide that stops moulds and yeasts.

**Dimethyl dicarbonate** (E242) acts as a preservative, especially for drinks, by disabling enzymes in microbes like yeasts, which can cause spoilage. It is now being added to wines in place of sulfur dioxide, to which some people are sensitive.

## 4. ENERGY DRINKS

*The Chemistry*

Sometimes we engage in exercise that makes a heavy demand on the body's reserves of energy and this demand can be met by taking a high energy drink with lots of caffeine, which allows the body's store of energy to be utilised. Sometimes our expenditure of energy also means we sweat a lot, and then we lose sodium and potassium so it may help to have these included in the drink. Vitamins that are involved in energy use may also be included.

### Relentless®

Some drinks promise more than mere energy and mineral replacement and this is one such product. It is a high caffeine energy drink produced by the Coca-Cola® company, and it comes in a 500 ml sized can, which provides 225 calories (from sugar) and 160 mg of caffeine, which is four times as much caffeine as in a 330 ml can of cola. (There are also 250 ml cans of Relentless.) The drink also supplies a few other ingredients for which there is some evidence of them boosting brain and body activity.

*Ingredients*

Carbonated water
Sugar
Citric acid
Taurine
Acidity regulator (sodium citrate)
Preservatives (potassium sorbate, sodium benzoate)
Colour (caramel E150d)
Flavourings
Caffeine
Vitamins (niacin, pantothenic acid, $B_6$, $B_{12}$)
Guarana

> **Carbonated water** is $CO_2$ dissolved under pressure in water.
> **Sugar** provides sweetness and calories.
> **Citric acid** provides the acid tang and the **acidity regulator** (E331) is its sodium salt, **sodium citrate**, which is added to keep the acid level stable.
> **Taurine** guards against oxidative stress due to over-strenuous exercise. It is needed by the body for several functions such as digestion, in blood vessels, the eyes, and white blood cells.

**Preservatives. Potassium sorbate** is E202 and is a highly effective fungicide that stops moulds and yeasts. Sorbates occur naturally in rowan berries. **Sodium benzoate** (E211) is a powerful antimicrobial that is particularly effective under acid conditions. It occurs naturally in cranberries.

**Colour (caramel E150d)** is modified caramel, which is more stable than normal caramel. It is manufactured by slowly heating sugar to around 180 °C with sulfur dioxide and ammonia.

**Flavourings** are not specified.

**Caffeine** is a stimulant, it makes breathing easier, and releases energy from stores within the body.

**Vitamins (niacin, pantothenic acid, $B_6$, $B_{12}$).** Niacin, pantothenic acid, and vitamin $B_6$ are involved in the release of energy from carbohydrates and fats. Vitamin $B_{12}$ is important for making red blood cells and the functioning of the nervous system.

**Guarana** comes from the seeds of the guarana plant, which grows in Brazil, and its seeds are a rich source of caffeine.

**Lucozade$^{®}$ Energy Original**

*Ingredients*

| | |
|---|---|
| Carbonated water | Preservatives (sodium benzoate, sodium |
| Glucose syrup (25%) | bisulfite) |
| Citric acid | Caffeine (0.012%) |
| Lactic acid | Antioxidant (ascorbic acid) |
| Flavouring | Colour (sunset yellow) |

**Carbonated water** is $CO_2$ dissolved under pressure in water.

**Glucose syrup (25%)** is made from maize (corn) starch and consists of linked glucose units.

**Citric acid** has a pleasing fruity tang.

**Lactic acid** acts as a preservative as well as providing some flavour.

**Flavouring** is not specified.

**Preservatives (sodium benzoate, sodium bisulfite).** The former is a powerful antimicrobial agent, especially under acid conditions; the latter is a chemical that kills microbes like fungi, yeasts, and bacteria.

**Caffeine (0.012%)** is a stimulant, it makes breathing easier, and releases energy from stores within the body. However, the amount here, in a typical 250 ml drink, is relatively insignificant.

**Antioxidant (ascorbic acid)** is vitamin C and has the food code E300.

**Colour (sunset yellow)** has food code E110.

## Quick Energy™

When all you want is something to make you feel more alert, then the traditional drink is coffee or tea. However, there is an alternative: a 'shot' of only 60 ml that is designed to pep you up, which contains no sugar and only 5 calories, while providing some B vitamins and amino acids. Here, the main ingredient is caffeine, which should have a rapid effect by releasing energy from the body's natural store of glycogen. Although it contains no sugar, it has three artificial sweeteners instead. It relies on releasing stored energy and it does this by stimulating functions within the body by means of caffeine, B vitamins, and amino acids.

*Ingredients*

| | |
|---|---|
| Caffeine | Preservatives (potassium sorbate, sodium benzoate) |
| Citric acid | Sweeteners (sucralose, acesulfame potassium, |
| Glucuronolactone |   aspartame) |
| Malic acid | Taurine |
| Niacin | Vitamin $B_6$ |
| Phenylalanine | Vitamin $B_{12}$ |

**Caffeine** is a stimulant, it makes breathing easier, and releases energy from stores within the body. There are around 150 mg or so in a 'shot'.

**Citric acid** has a pleasing fruity tang.

**Glucuronolactone** is a natural chemical made by the liver from glucose and is needed by the body's connective tissue and tendons. It is also said to improve mental alertness and counteract tiredness.

**Malic acid** is the acid of unripe apples. Here is it acts as an antioxidant.

**Niacin** is vitamin $B_3$, and also known as nicotinic acid. It assists enzymes in releasing energy from carbohydrate.

**Phenylalanine** is an essential amino acid for humans. However, there are people who suffer from the genetic disorder phenylketonuria and who convert it to another chemical – a phenyl ketone that can cause brain damage. These people should avoid drinking Quick Energy.

**Preservatives (potassium sorbate** and **sodium benzoate)** are there to prevent germs from colonising the product.

**Sweeteners** are of three kinds: **sucralose, acesulfame potassium**, and **aspartame.**

**Taurine** is needed by the body for several functions such as in digestion, in the functioning of blood vessels, in the eyes, and in white blood cells. In Quick Energy its role is to improve muscle tone.

**Vitamin B$_6$** (also known as pyridoxine) is needed to convert amino acids into the protein of muscles and enzymes. Another key role is in the release of energy from carbohydrates.

**Vitamin B$_{12}$** is important for making red blood cells and the functioning of the nervous system.

## 5.   HOT DRINKS

*The Chemistry*

Many people say nothing cheers like a cup of tea, and its benefit is not simply a supply of caffeine and theobromine, which are chemically similar. Tea contains small amounts of many other natural chemicals, which might account for the belief that nothing quite satisfies in the same way as a cup of tea. Coffee is also a caffeine-based drink, which provides a wider and stronger range of flavours.

Other hot drinks are supposed to comfort the drinker and these are cocoa, hot chocolate, Ovaltine®, and Horlicks®. These generally provide lots of calories but the following offers the possibility of enjoying the cocoa taste without many calories.

© Shutterstock

## Cadbury® Highlights

This version provides only 40 calories per cup as opposed to the 165 calories that the normal version provides.

*Ingredients*

| | |
|---|---|
| Fat-reduced cocoa | Maltodextrin |
| Whey, dried | Salt |
| Skimmed milk, dried | Thickener E407 |
| Glucose syrup | Anticaking agent E551 |
| Vegetable oil | Milk proteins |
| Milk chocolate (6%) | Aspartame |
|   made of milk, sugar, | Acesulfame K |
|   cocoa butter, cocoa | Flavourings |
|   mass, soya lecithin | Emulsifier E471 |

**Fat-reduced cocoa** is the 45% of the cocoa bean that remains after it has been ground up and hot pressed to remove the cocoa butter.

**Whey** is the liquid that remains when milk has been curdled, as happens in the making of cheese. The curds are then filtered off and the whey spray-dried.

**Skimmed milk, dried** is milk without its cream, which has been spray-dried to form powder.

**Glucose syrup** is produced by the action of enzymes on starch. These break its long chains down into smaller units.

**Vegetable oil** is produced by plants and can be sourced from olives, rape, and sunflowers.

**Milk chocolate (6%) made of milk, sugar, cocoa butter and cocoa mass.** As its name implies, this is like the traditional dairy milk chocolate. **Lecithin** (E322) is an emulsifier, and the one used here is extracted from soya.

**Maltodextrin** consists of short chains of linked glucose molecules, and it is made from wheat starch.

**Thickener E407** is also known as carrageenan and is used to give the finished drink the right smoothness.

**Anticaking agent E551** is silicon dioxide and is there to prevent the powder sticking together during storage, so that it remains free-flowing and easily disperses throughout the drink when boiling water is added.

**Milk proteins** consist of chains of linked amino acids. There are 22 different kinds of amino acid, some of which are essential to our diet and which our bodies cannot make for itself, but milk contains all of them.

**Aspartame** and **acesulfame K** are intense, calorie-free, sweeteners that together intensify each other's effect, so less of each is needed.

**Flavourings** are not specified.

**Emulsifier E471** is mono- and di-glycosides of fatty acids and ensures that all the ingredients are nicely blended together in the final drink.

## Coffee-mate®

This product is added to coffee to give it a sweet and creamy taste without cooling it down too much – rather like adding hot milk.

*Ingredients*

Glucose syrup
Palm oil
Milk proteins
Stabilisers (sodium hexametaphosphate, sodium citrate)
Acidity regulator (dipotassium phosphate)
Emulsifiers (Mono- and di-glyceride esters of fatty acids)
Mono- and di-acetyl tartaric acid ester of mono- and di-glycerides of fatty acids)
Trisodium citrate
Anti-caking agent (silicon dioxide)
Colour (riboflavin)

**Glucose syrup** is produced by the action of enzymes on starch. These break its long chains down into smaller units.

**Palm oil** is and edible oil extracted from the pulp of the fruit of oil palm plants.

**Milk proteins** consist of chains of linked amino acids. There are 22 different kinds of amino acid, some of which are essential to our diet and which our bodies cannot make for itself, but milk contains all of them.

**Stabilisers (sodium hexametaphosphate, sodium citrate)** the former has the code E452 and the latter is E331. The former acts to sequester interfering anions; the latter prevents the product from coagulating into lumps when hot water is added.

**Acidity regulator (dipotassium phosphate)** has the code E340 and ensures the product is stable during storage.

**Emulsifiers (Mono- and di-glyceride esters of fatty acids)** have the code E471 and act to ensure that when Coffee-mate® is added to coffee its ingredients are dispersed smoothly. If you want whipped coffee these will produce stable foam.

**Mono- and di-acetyl tartaric acid ester of mono- and di-glycerides of fatty acids** has the code E472e and also acts an as emulsifier.

**Trisodium citrate** has the code E331 and is an acidity regulator.

**Anti-caking agent (silicon dioxide)** has the code E551 and it acts to keep the product free-flowing.

**Colour (riboflavin)** is vitamin $B_2$ and which has the code E101. It is orange coloured.

## 6.  DESSERTS

*The Chemistry*

It is always nice to end a meal with something sweet, but with sweetness often comes sugar and calories. The calories may be in the form of glucose (also known as dextrose), fructose, and honey. Here we look at some that you can enjoy without feeling guilty, and then at the most popular of all, ice cream, in both its normal form and in a lower-calorie version.

### Shape® Strawberry Yoghurt

This is a product in a range of low calorie desserts, and it also claims to make you feel fuller for longer. It does this by having a higher level of fibre, which the digestive system cannot break down so it remains in the stomach for longer.

*Ingredients*

Skimmed milk
Fruit (6%): strawberry
Skimmed milk powder
Milk powder
Fibre: guar gum, oligofructose
Stabilisers: modified maize starch,
  carrageen
Flavouring
Acidity regulators: sodium citrate,
  citric acid
Yogurt cultures
Colour: cochineal
Sweeteners: aspartame, acesulfame K

Some ingredients are self-explanatory, namely **skimmed milk, fruit (6%) strawberry, skimmed milk powder, milk powder,** and **yogurt cultures.**

**Fibre: guar gum** is extracted from the beans of the guar plant. It consists of long chains of mannose molecules with galactose molecules attached. These are the carbohydrates that resist being digested. The other carbohydrate is **oligofructose**, which can be extracted from chicory or made from sugar. It also has a creamy 'mouth-feel.'

**Stabilisers** prevent the other ingredients from separating so you don't get solids at the bottom of the yoghurt carton and a liquid layer on top. Here they are **modified maize starch**, also known as cornflour, and **carrageen**, which is extracted from seaweed and is a galactose polymer. Carrageen also acts as a thickener.

**Flavouring** is no doubt that of strawberries which comes mainly from 4-hydroxy-2,5-dimethyl-3-furnanone.

**Acidity regulators: sodium citrate, citric acid** (E331, E330) together stabilise the pH of the yogurt to keep it in the optimum range of between pH4 and pH5.

**Yogurt cultures** are bacteria and they are *Lactobacillus bulgaricus* and *Streptococcus thermophiles.* (*Lactobacillus bulgaricus* is now known as *Lactobacillus delbrueckii* subspecies *Bulgaricus.*)

**Colour: cochineal** is obtained from the female scale insect (*Dactylopium coccus*), which lives on cacti. The chemical is red carminic acid.

**Sweeteners: aspartame and acesulfame K** are intense, calorie-free, sweeteners that together intensify each other's effect, so less of each is needed.

### Hartley's® Sugar Free Jelly

These jellies contain fewer than 10 calories per serving while having a fruity and sweet taste. The following are the ingredients in their lemon and lime variety.

*Ingredients*

Gelatine
Adipic acid
Flavourings
Acidity regulator: trisodium citrate
Sweeteners:
 aspartame
 acesulfame K

Fumaric acid
Colours:
 copper complexes of chlorophylls
 mixed carotenes

**Gelatine** gives the jelly its semi-solid structure.

**Adipic acid** (E355) is used partly for its fruity flavour and partly because it helps the jelly to set.

**Flavourings** are not specified but are lime oil and lemon oil. These are a complex mixture of many molecules.

Lime oil contains the hydrocarbons alpha-pinene, beta-pinene, myrcene, limonene, terpinolene, and 1,8-ceneole, the alcohols linalool and borneol, and the ketone citral.

Lemon oil also contains alpha-pinene, beta-pinene, myrcene, and limonene, along with the hydrocarbons camphene, sabinene, alpha-terpinene, beta-bisabolenethe, *trans*-alpha-bergamotene and lemonal, together with linalool and nerol.

**Acidity regulator: trisodium citrate.** This has the food code E331 and it keeps the acidity of the jelly constant.

**Sweeteners: aspartame** and **acesulfame K** are intense, calorie-free, sweeteners that together intensify each other's effect, so less of each is needed.

**Fumaric acid** (E297) has a fruity flavour and also helps keeps the acid level of the jelly stable.

**Colours: copper complexes of chlorophylls** (E141) and **mixed carotenes** (E160a) together provide the pale green/yellow colour. These are natural colourants and part of the trend to replace dyes with natural colourants.

## Ice Cream

*The Chemistry*

The name of this dessert is a misnomer: it isn't frozen cream. One of the most popular brands of vanilla ice cream is Carte D'Or® and it comes in two varieties: normal and light. The normal version has the following ingredients, all of which are naturally derived.

## Carte D'Or® Ice Cream

*Ingredients*

Reconstituted skimmed milk
Glucose–fructose syrup
Sugar
Glucose syrup
Coconut oil
Whey solids (milk)
Stabilisers (locust bean gum, guar gum, carrageenan)

Emulsifier (mono- and di-glycerides of fatty acids)
Vanilla bean pieces
Natural vanilla extract from Madagascar
Colour (carotenes)

Most of these ingredients are self-explanatory, such as **sugar, coconut oil, vanilla bean pieces** and **natural vanilla extract**.

**Reconstituted skimmed milk** is skimmed-milk powder with water added.

**Glucose–fructose syrup** consists of the same chemical units, fructose and glucose, that make up sugar, but with more of the sweeter fructose than of glucose. Chemically it is rather like honey. It can be made from corn (maize) starch.

**Glucose syrup** is made from maize (corn) starch and like all starch it consists entirely of linked glucose units.

**Whey solids (milk)** are produced by spray-drying the watery liquid left behind when milk is curdled.

**Stabilisers (locust bean gum, guar gum, carrageenan)** are natural gums that are used as thickening agents. The locust bean gum (E410) comes from the seeds of the carob tree, guar gum (E412) comes from guar seeds, and carrageenan (E407) is obtained from red seaweed.

**Emulsifier (mono- and di-glycerides of fatty acids)** have the food code E471. They ensure that the texture of the ice cream remains consistent during storage.

**Colour (carotenes)** have the food code E160a and they are the orange-coloured molecules that give carrots and sweet potatoes their colour.

A family sized tub (900 ml) of the above ice cream provides 900 calories, a lot of which come in the form of sugar and similar carbohydrates.

There are many brands of ice cream such as Wall's, Häagen-Dazs®, Ben and Jerry's®, plus own-brand supermarket ones.

## Carte D'Or® Light

This version of Carte D'Or® has replaced some of the calories of the former product, not with artificial sweeteners, but with water and as a result the carton contains 630 calories, which is 30% less than the standard version.

### Ingredients

Reconstituted skimmed milk
Water
Sugar
Glucose syrup
Oligofructose syrup
Coconut oil
Whey solids (milk)
Stabilisers (locust bean gum, guar
  gum, carrageenan)
Emulsifier (mono- and di-glycerides
  of fatty acids)
Natural vanilla flavouring from
  Madagascar
Vanilla bean pieces
Colours (mixed carotenes)

Most of these ingredients are self-explanatory, such as **sugar, coconut oil, vanilla bean pieces** and **natural vanilla flavouring**.

**Reconstituted skimmed milk** is skimmed-milk powder with water added.

**Water** is now the second most important ingredient.

**Glucose syrup** is made from maize (corn) starch and like all starch it consists entirely of linked glucose units.

**Oligofructose syrup** is a carbohydrate consisting of a short chain of linked fructose units.

**Whey solids (milk)** is produced by spray-drying the watery layer left behind when milk is curdled.

**Stabilisers (locust bean gum, guar gum, carrageenan)** are natural gums that are used as thickening agents. The locust bean gum (E410) comes from the seeds of the carob tree, guar gum (E412) comes from guar seeds, and carrageenan (E407) is obtained from red seaweed.

**Emulsifier (mono- and di-glycerides of fatty acids)** have the food code E471. They ensure that the texture of the ice cream remains consistent during storage.

**Colours** (mixed carotenes) have the food code E160a and they are orange coloured. Carotenes are the natural colour of carrots.

## 7. VITAMIN SUPPLEMENTS

*The Chemistry*

A balanced diet should provide all the essential nutrients (proteins, fats, and carbohydrates) and micronutrients (vitamins and minerals) that the body needs. It is most unlikely that anyone today will suffer a deficiency of vitamins and minerals that previous generations were prone to, and which caused conditions like pellagra (lack of niacin), scurvy (lack of vitamin C) and rickets (lack of vitamin D). Lack of niacin causes large, weeping sores and lack of energy. Pellegra was widespread among prisoners in camps where the diet failed to provide any fish, poultry, or nuts, which are rich sources. Scurvy afflicted sailors who lacked fresh fruit and vegetables for weeks on end. Rickets is again becoming a recognised problem in the UK population.

When more of a micronutrient is provided than the body needs, it can be stored against leaner times – although this is not possible with some vitamins, such as the all-important vitamin C. Even a vitamin that can be stored may eventually run out, as can happen with vitamin D. A diet that includes five pieces or portions of fruit or vegetables a day should ensure

© Shutterstock

enough vitamin C. Vitamin D cannot be catered for this way because it is a fat-soluble vitamin, so the diet should include fish oil, eggs, or margarine to supplement the amount that is formed when the skin is exposed to sunlight. Even so, it is possible to guard against dietary deficiency by taking a pill that provides some of those that are needed, such as the following product.

## Berocca® Vitamins and Minerals

On its packaging, Berocca® says that it contains 'tailored vitamins and minerals to set you up for a really good day.' These ingredients are discussed here under the headings vitamins and minerals. This is the fizzy tablet version. There is also a capsule version.

*Contents*

Citric acid
Sodium hydrogen carbonate
Vitamin C
Magnesium sulphate
Mannitol
Calcium carbonate
Magnesium carbonate
Flavouring
Sodium carbonate
Niacin
Sweeteners (Acesulfame K, Aspartame)
Salt
Zinc citrate
Colour (Beetroot red, Beta carotene)
Pantothenic acid

Maltodextrin
Riboflavin
Thiamin
Acacia gum
Vitamin B$_6$
Partially hydrogenated soybean oil
Sugar
Trisodium citrate
Antifoaming agent (Polysorbate 60)
Folacin (folic acid)
Antioxidants (Alpha-tocopherol, Sodium ascorbate)
Biotin
Vitamin B$_{12}$

**Citric acid** and **sodium hydrogen carbonate** (formerly known as bicarbonate of soda) are the two ingredients that react chemically when added to water, and the tablet fizzes as it generates bubbles of carbon dioxide.

*Vitamins*

**Vitamin C** (ascorbic acid) is something the body cannot store and we need around 60 mg a day to remain healthy. Each Berocca® tablet provides eight times this amount.
**Niacin** is needed to prevent pellagra.

**Pantothenic acid** is essential to energy production and our ability to cope with stress.

**Riboflavin** (vitamin $B_2$) is important for keeping the skin and eyes in good condition, as well as being essential in foetal development.

**Thiamin** (vitamin $B_1$) is needed to prevent beriberi whose symptoms are weakness, pain, and a swelling of the body.

**Vitamin $B_6$** is essential for making the amino acids that form proteins and it is necessary in order to release the glucose stored by the body as its source of instant energy.

**Folacin (folic acid)** is especially important for a woman seeking to have a baby. If she lacks this vitamin when she conceives, then there is a high risk of her baby being born with spina bifida (exposed spine).

**Biotin** is vital to the skin, hair, nerves, and sex organs. Intestinal bacteria provide most, but not all, of the 150 micrograms a day that we need.

**Vitamin $B_{12}$.** When this is lacking it leads to pernicious anaemia. We need only 1.5 micrograms of this a day and Berocca provides almost a week's supply.

**Alpha-tocopherol** is vitamin E, which is an antioxidant.

**Sodium ascorbate** is a form of vitamin C.

*Minerals*

**Magnesium** is essential for the smooth operation of muscles and enzymes.

**Calcium carbonate** may also help produce the fizz, but this metal is needed, and a lot of the calcium we absorb from our diet ends up in our bones but this metal also has a role in cells, blood clotting, and hormones.

**Zinc** is needed for enzymes and it is essential for the sex glands, especially the prostate. Lack of zinc in the diet leads to poor sexual development.

*Other Ingredients*

**Mannitol** is a sweetening agent, but does not cause tooth-decay.

**Salt** acts as a preservative.

**Acacia gum** acts to glue the ingredients together in the tablet.

**Trisodium citrate** is an acidity regulator to ensure the final drink has the right pH.

**Sodium carbonate** may also help produce the fizz, but its main purpose is to act as a filler.

**Acesulfame K, aspartame** and **sugar** are there to counteract the bitter taste of other ingredients.

**Maltodextrin** acts as a binder.

**Partially hydrogenated soybean oil** is less prone to oxidation, which can form molecules with a rancid smell and taste, but less so when the oil has been hydrogenated. This ingredient is needed because alpha-tocopherol (vitamin E) is oil-soluble and comes in this form.

**Polysorbate 60** has both water-seeking and oil-seeking components in the molecule and as such acts to ensure all the ingredients dissolve in water.

## Boots® Selenium with Vitamins A, C, and E

*The Chemistry*

There are some vitamin supplements that have been formulated to target certain threats, such as those from free radicals. Free radicals are particularly dangerous and every cell of the body has to contend with them. These molecules have an extra electron which makes them very reactive and chemically aggressive. They are able to react with almost all other molecules, leading to the formation of unwanted materials and can convert useful molecules to ones that are useless. The cell has its own natural defences like vitamin E. The following product is designed to strengthen the body's supplies of this vitamin.

It also contains selenium, which is an element that is essential for the human body as a vital part of certain enzymes, including those of the thyroid gland. It is especially important for men, who need selenium to make sperm. Excess selenium is toxic but in moderate amounts it may protect the body against other toxic metals like mercury. The *maximum* daily intake should not exceed 0.5 mg, and normal intake need be only a tenth of this (0.05 mg). A dose of 5 mg would product toxic effects, the most noticeable of which is foul smelling breath lasting many weeks.

*Contents*

Sodium selenite
Retinyl palmitate
DL-alpha tocopheryl acetate
Soya bean oil
Ascorbic acid
Glycerol

Soya lecithin
Gelatin
Beeswax
Silicon dioxide
Iron oxide

**Sodium selenite** provides selenium, which is a vital part of certain enzymes, including those of the thyroid gland. Each capsule contains 50 micrograms (0.05 mg) of selenium as sodium selenite.

*Vitamins*

**Retinyl palmitate** is a form of vitamin A, and an antioxidant. The palmitate comes from palmitic acid, the major component of palm oil.

**DL-alpha tocopheryl acetate** is vitamin E and is an antioxidant.

**Soya bean oil** is required because vitamins A and E are oil-soluble. Soya bean oil is rich in polyunsaturated linoleic acid and monounsaturated oleic fatty acids.

**Ascorbic acid** is vitamin C, which is water soluble.

**Glycerol** is there to solubilise the selenium sulphite and the vitamin C.

**Soya lecithin** is an emulsifier able to make the mixture of oil and other ingredients blend together.

The capsules are made of **gelatin**.

**Beeswax** is the glazing agent to give the capsules a shine.

**Silicon dioxide** (E551) provides the coating for the capsules.

**Iron oxide** colours the capsule coating dark brown.

## Boots® Cod Liver Oil and Calcium

*The Chemistry*

As we get older, bones get weaker. Bones are mainly calcium phosphate and they can act as a reserve supply of calcium, which the body needs for various processes, such as creating new cells and controlling the pH of body fluids. Calcium in the bones is continually being released and replaced, but as a person ages the latter process no longer quite compensates for the former, and bones get weaker and are prone to break easily. In order to prevent this, the diet should include some calcium-rich foods like sardines, eggs, and cheese. The following product provides two kinds of essential nutrients: omega-3 oil and calcium, which together might prevent bone weakening.

*Contents*

Calcium carbonate
Cod liver oil
Hydrogenated soya bean oil
Glycerol
Sorbitol
DL-alpha tocopheryl acetate

Colours: titanium dioxide, iron oxide
Retinyl palmitate
Antioxidant: DL-alpha tocopherol
Fractionated coconut oil
Phytomenadione
Cholecalciferol

**Calcium carbonate** is the mineral of limestone, although that which is used in these capsules is pure and made by the chemical industry.

**Cod liver oil** is a rich source of omega-3 oil.

**Hydrogenated soya bean oil** has been reacted with hydrogen thereby making its fatty acid components saturated and less likely to be oxidised by the oxygen of the air, which would make it rancid.

**Glycerol** is added as a thickening agent, and it helps to solubilise the other ingredients.

**Sorbitol** acts as a lubricant and is non-greasy.

**DL-alpha tocopheryl acetate** is vitamin E and is also an antioxidant.

**Colours: titanium dioxide** (white, E171) and **iron oxide** (red, E172) are used to colour the capsule's outer covering.

**Retinyl palmitate** is a form of vitamin A, and an antioxidant.

**Antioxidant: DL-alpha tocopherol** (E307) is vitamin E and is an antioxidant.

**Fractionated coconut oil** acts as a stable medium for the added vitamins.

**Phytomenadione** is a fat-soluble vitamin, sometimes called vitamin K. This, in conjunction with **cholecalciferol**, which is a form of vitamin D, has been shown to improve bone density when taken in conjunction with calcium.

CHAPTER 10

# The Living Room

Here's where we can relax in an evening, enjoy some tasty snacks, have a gin and tonic, watch the TV, while the children quietly engage in creative play. Of course things are rarely as enjoyable as this, but whatever we do to relax, it will still involve chemistry. There are scores of snacks and hundreds of drinks to

Chemistry at Home: Exploring the Ingredients in Everyday Products
By John Emsley
© John Emsley, 2015
Published by the Royal Society of Chemistry, www.rsc.org

choose from. The ones I've chosen for this chapter reflect my own preferences, and they are meant more to show the role that chemistry plays even in this haven of family life.

1. Snacks (Pringles® Sour Cream and Onion Flavour, Wotsits™, Alpen® Light Summer Fruits Bar, Duchy Originals® Organic Lemon Shortbread)
2. Remote control (Energizer® Lithium Batteries)
3. Drinks (Ribena Light®, Bombay Sapphire™ Gin, Schweppes® Indian Tonic Water, Schweppes® Slimline Tonic Water)
4. Toys (Play-Doh®)

## 1. SNACKS

*The Chemistry*

As in the previous chapter, the chemistry of what you are eating is less important than the amount of calories you are consuming, and often these are surplus to your daily requirements. Nibble a small bowl of roasted peanuts and you may well consume 200 calories because a single peanut weighing a mere 1 gram provides 6 calories. And a bag of traditional crisps is likely to provide about the same amount, although some crisps are now sold in bags of only 25 grams. Of course foods that are fried absorb fat and that is the main source of their calories. Fat provides 9 calories per gram. The only way in which such savoury snacks have been made more acceptable has been through the use of less salt to flavour them.

Although chemists have discovered sweeteners that can mimic the taste of sugars and replace them, thereby saving calories, their attempts to produce substitutes for fats have been much less successful, and some that were introduced, such as Simplesse® in the 1990s, were found to be prone to cause diarrhoea and never caught on.

There are many snacks on the market – almost too many to mention – some are fried, some are baked, some are made of potato, some are made of flour. Among them are the popular Walkers™ crisps, Seabrook® crisps, Jacob's™ Mini Cheddars®, Hula Hoops®, Jacob's™ Twiglets®, Popchips™, and Doritos®. Here we look at just two such snacks.

## Pringles® Sour Cream and Onion Flavour

*The Chemistry*

These snacks come in a tube, weigh 190 grams, and are assumed to represent around eight portions, each of which would provide around 120 calories. However, you might easily share a tube of Pringles® between a family of four, in which case you would be taking in around 240  calories each. There are about a dozen flavour varieties of Pringles®.

*Ingredients*

| | |
|---|---|
| Dehydrated potato | Monosodium glutamate |
| Vegetable oil | Disodium inosinate |
| Vegetable fat | Onion powder |
| Rice flour | Dextrose |
| Wheat starch | Sweet whey powder |
| Maltodextrin | Lactose from milk |
| Salt | Citric acid |
| Sour cream and onion flavour | Lactic acid |
| The components of this are listed | Malic acid |
| as follows: | Emulsifier E471 |

The first five ingredients listed above are self-explanatory.

**Maltodextrin** is made from starch and consists of joined-up glucose molecules with as many as 17 in a chain. Maltodextrin tastes slightly sweet.

**Salt** is still needed to give the product its familiar  taste.

**Sour cream and onion flavour** consists of the following:

**Monosodium glutamate** (E621) and **disodium inosinate** (E631) not only improve the umami flavour but can partly replace salt.

**Onion powder** is essentially dried onion. When an onion is cut, enzymes get to work on the molecule 1-propenyl cysteine sulf-oxide, which they covert to thiopropionaldehyde-$S$-oxide, and this is the molecule that gives onion its characteristic flavour.

**Dextrose** is another name for glucose.

**Sweet whey powder** is mainly lactose and is produced by spray-drying the liquid left after milk has been curdled.

**Lactose** consists of two linked carbohydrate molecules, glucose and galactose, bonded together.

**Citric acid** (E330) is the acid of citrus fruits, **lactic acid** (E270) is also known as milk acid, and **malic acid** (E296) is found in apples, grapes, and other fruits. These act as preservatives.

**Emulsifier E471** consists of long chain fatty acids bonded to glycerol. It ensures that the various ingredients blend together well.

Some snacks are baked rather than fried, and they provide the same satisfying crunch without the same number of excess calories.

**Walkers Wotsits™**

These cheese-flavoured snacks provide only 96 calories, although a bag contains only 20 grams. They are made from maize flour and sunflower oil with a few other ingredients.

*Ingredients*

Maize
Sunflower oil (33%)
Cheese flavour, cheese powder 7%
Lactose (from milk)
Flavour enhancer (disodium
  5'-ribonucleotide)

Colours (paprika extract, annatto)
Salt
Potassium chloride

The packet warns that the product may contain traces of celery, barley, gluten, wheat, mustard, and soya, because products containing these ingredients are made in the same factory.

**Maize** flour has no gluten, whereas wheat flour does. This makes Wotsits™ a possible snack for those who suffer gluten intolerance.

**Sunflower oil (33%)** consists of 30% unsaturated oleic fatty acid, 60% polyunsaturated linoleic acid, and as such is regarded as healthier than animal fats.

**Cheese flavour, cheese powder 7%** is what it says.

**Lactose (from milk)** is a carbohydrate that reinforces the cheese flavour. Some people are, however, lactose intolerant.

**Flavour enhancer (disodium 5'-ribonucleotide)** is added to increase the umami aspect of the flavour. It has the food code E635 and is made by combining the sodium salts of two naturally occurring compounds, inosinic acid and guanylic acid.

**Colours (paprika extract,** food code **E160c, annatto,** food code **E160b)** are natural red colourants.

**Salt** and **potassium chloride** provide the necessary salty taste without this relying on salt alone.

## Alpen® Light Summer Fruits Bar

*The Chemistry*

This is a tasty snack bar that does not provide much in the way of sugar, fats, or salt and it contains fewer than 70 calories per bar. (Regular Alpen® bars deliver around 120 calories.) Most of its ingredients are naturally sourced, so what follows is an explanation of the less familiar ones.

*Ingredients*

| | |
|---|---|
| Cereals (wheat, oats, rice) 42% | Milk lactose |
| Oligofructose 30% | Milk yogurt powder |
| Apple pieces and apple juice concentrate 6.5% | Salt |
| Strawberries, raspberries, and cranberries 3.5% | Maize starch |
| Sugar | Flavourings |
| Glycerol | Skimmed milk powder |
| Sulphur dioxide | Milk whey powder |
| Glucose syrup | Defatted wheat germ |
| Vegetable oil | Citric acid |
| Dextrose | Lecithin |
| Wheat gluten | Malic acid |
| Malted barley extract | Tocopherols |

The fruit ingredients are listed and these are dried fruit, and in the case of apple there is also apple juice concentrate.

Other ingredients which need no explaining are sugar, glucose syrup, dried yogurt, dried skimmed milk and dried milk whey.

**Oligofructose (30%)** is a carbohydrate consisting of a short chain of linked fructose units.

**Glycerol** (food code E422) is the humectant and there to keep the bar moist. It is slightly sweet tasting.

**Sulphur dioxide** (food code E220) acts as a preservative for the fruits used to make the Alpen® bar and, as such, tiny residues of this are present in the product.

**Vegetable oil** is not specified but can be one of various edible oils, such as sunflower, linseed, palm, or rapeseed oil.

**Dextrose** is another name for the glucose used in processed foods.

**Wheat gluten** is listed so that those who are affected by this are informed.

**Malted barley extract** is also known just as 'malt' and is a thick syrup made by extracting malted barley with water and then concentrating this by heating under vacuum to remove almost all of the water.

**Milk lactose** is a carbohydrate that is extracted from the whey that is a by-product of cheese making.

**Citric acid** acts as a preservative and it has a pleasing fruity acid tang.

**Lecithin** (food code E322) is an emulsifier and the one used here is extracted from soya.

**Malic acid** (food code E296) regulates the overall acid level within the bar and acts as a preservative.

**Tocopherols** (food code E306) are a form of vitamin E and a powerful antioxidant.

Other fruit bars are Jordans®, Nature Valley®, Special K®, Weetabix®, Nãkd®, *etc.*

Alternatively, you might prefer a biscuit when drinking a cup of tea and there are scores of these to choose from.

## Duchy Originals® Organic Lemon Shortbread

These all-butter shortbread biscuits use British wheat flour, including flour from wheat grown on Duchy Home Farm at Highgrove in Gloucestershire, the home of Prince Charles. It also includes crystallised Sicilian lemon peel.

*Ingredients*

Wheat flour
Butter (27%)
Sugar
Crystallised Sicilian lemon peel (7%)

Sodium bicarbonate
Ammonium bicarbonate
Sea salt

**Wheat flour**, some of which comes from Duchy Home Farm, which is a certified organic farm and as such eschews the use of modern fertilizers and pesticides, preferring instead the first generation chemicals that farmers used before the advent of the agrochemical industry.

**Butter** (27%) source unspecified, but presumably from an organic farm.

**Sugar** source unspecified, but presumably from an organic farm.

**Crystallised Sicilian lemon peel** (7%) consists of lemon peel, corn syrup, sugar, and concentrated lemon juice. Presumably the lemons were grown on an organic farm in Sicily.

**Sicilian lemon oil** is presumably from an organic lemon grove in Sicily.

**Sodium bicarbonate** (E500) and **ammonium bicarbonate** (E503) are said to be 'approved non-organic' chemicals that act as raising agents – see the section on baking powder in Chapter 8, The Kitchen.

**Sea salt** is sodium chloride.

There are other shortbread biscuits such as Walker's® and Crawford's®.

## 2.  REMOTE CONTROL

### Energizer® Lithium Batteries

*The Chemistry*

Many appliances, such as tablets and phones, come with built-in rechargeable lithium batteries. We also need long-lasting lithium batteries (AA and AAA sizes) for powering things in the living room, such as clocks and the remote control. So how do those batteries work?

Electric current is generated by chemical changes inside the battery, and these occur at the anode and the cathode and involve a transfer of electrons. The chemical reaction begins when the battery becomes part of an external circuit through which the electrons can flow from one electrode to the other.

Lithium batteries consist of a positive electrode (anode) made of lithium and a negative electrode (cathode) made of manganese dioxide and graphite, between which is an electrolyte consisting of lithium perchlorate in a mixed solvent of propylene carbonate and dimethoxyethane. Manganese dioxide ($MnO_2$) has been used in dry-cell batteries from the earliest days. Lithium atoms release electrons to become positive ions ($Li^+$), the

© Shutterstock

manganese ions gain them, going thereby from manganese $Mn^{4+}$ to $Mn^{3+}$.

The electrolyte is the salt lithium perchlorate ($LiClO_4$) dissolved in a solvent and which provides a conducting medium between the anode and cathode. The solvent has to dissolve the $Li^+$ and $ClO_4^-$, but the usual solvent for these, water, cannot be used because this would react violently with the lithium metal. The solvent here is propylene but by itself this is a very viscous liquid so it is diluted with dimethoxyethane.

The other major brands of lithium battery are Duracell® and Panasonic®. The smaller type of coin-sized long-life batteries are also needed for some small appliances.

## 3.   DRINKS

*The Chemistry*

It is pleasant to relax in an evening and even pleasanter when you refresh yourself with a drink. This does not need to be an alcoholic one, nor need it be sugary. The drinks in this section are designed just to refresh and or to relax you, ideally providing as few sugar calories as possible. (Fizzy, non-alcoholic drinks are included in Chapter 9.)

### Ribena Light®

This drink has been reformulated to boost its vitamin C content and reduce its calories by replacing sugar with sweeteners.

*Ingredients*

| | |
|---|---|
| Water | Vitamin C |
| Blackcurrant juice from concentrate (7%) | Sweeteners (aspartame, acesulfame K) |
| | Citric acid |
| Malic acid | Stabiliser (xanthan gum) |
| Acidity regulators (calcium hydroxide, calcium carbonate) | Flavouring |
| | Colour (anthrocyanins) |

**Blackcurrant juice from concentrate (7%)** is the natural product but which has been processed and whose vitamin C content will thereby have been depleted.

**Malic acid** (E296) is the sour taste of unripe fruits and acts as a preservative.

**Acidity regulators (calcium hydroxide** food code E526, and **calcium carbonate** food code E170) are used to control the pH of the drink so that it remains constant.

**Vitamin C** in this drink will provide the 60 mg that is a daily requirement for adults and children, although the minimum intake needed to prevent vitamin C deficiency is 40 mg.

**Aspartame** (food code E951) and **acesulfame K** (food code E950) are intense, calorie-free sweeteners that together intensify each other's effect, so less of each is needed.

**Citric acid** (food code E330) is a natural acid produced by many fruits and especially oranges and lemons. It also acts as a preservative.

**Xanthan gum** (food code E415) protects and preserves the blackcurrant juice concentrate.

**Anthrocyanins** (food code E163) are red–purple pigments and the natural colour of blackcurrants.

Other makes of fruit drink are available under brand names such as Robinsons®, Ocean Spray®, Innocent® and supermarket own brands.

## Gin

*The Chemistry*

This particular spirit, historically linked to London, has had a revival this century with a large number of artisan gins now on the market, and the number of varieties now exceeds 150. Some have whimsical names such as Bathtub™, Tanqueray No.10®, and Ish®.

The basis of gin is the alcohol distilled from fermented grain. (Vodka is traditionally made from potatoes.) This is infused with natural flavouring agents, referred to as botanicals, and then distilled for a second time. The distillate now contains traces of volatile chemicals, released from the botanicals, and it is these which give gin its flavour and aroma. The liquor is then diluted with pure water to give the desired strength – be it 37.5%, which is the minimum amount permitted if it is to be called gin, 40%, or even 50%.

Unlike most other foods, alcoholic drinks are not required to carry a list of ingredients or contents.

The following is one of the more popular 40% gins, and its botanicals are typical of many other gins, although the exact proportions used by the distillery are what make an individual brand distinguishable.

## Bombay Sapphire™ Gin

This premium London dry gin is used to be distilled in Warrington, Lancashire but is now made in Leverstoke in Hampshire. The use of the term 'London' is a guarantee that the gin has been made in the way described above and that its methanol content is less than 5 grams per 100 litres (0.005%). Methanol is produced during fermentation and is present in small amounts in all drinks, including wines, but being more volatile than ethanol, it is easily removed from spirits at the start of the distillation process.

*Ingredients*

| | |
|---|---|
| Alcohol (40%) | Liquorice |
| Botanicals: | Orris |
|   Juniper berries | Coriander |
|   Angelica | Cassia bark |
|   Almonds | Cubeb berries |
|   Lemon peel | Grains of paradise |

**Juniper berries** are the basis of all gins. They release oil consisting of cadinene, camphene, terpineol, and alpha-pinene. Cadinene provides the most recognisable aroma of gin; camphene has the smell of camphor; terpineol has a floral, lilac, aroma; alpha-pinene smells of pine leaves in its pure state.

**Angelica** root is also an essential part of the botanicals. The main volatile components it releases are 3-butylidene-4,5-dihydrophthalide, also known as lingustilide, and beta-terebangelene, both with a sweet aroma.

**Almonds** contain benzaldehyde, which has the characteristic aroma we associate with these nuts, and this is released on heating.

**Lemon peel** releases several molecules, of which 2,6-dimethyl-5-heptenal is the main one. In the pure state this has a fruity melon-like flavour.

**Liquorice** contains glycyrrhizic acid, also known as glycyrrhizin. In its pure state this is 30 times sweeter than sugar.

**Orris** provides cetonal, a woody note with a hint of raspberry.

**Coriander** seeds add linalool to the flavour profile; it has a sweet floral aroma.

**Cassia bark**'s main flavour component is cinnamaldehyde, which has the spicy flavour of cinnamon itself.

**Cubeb berries** provide various oils, of which sabinene and cubebene predominate. The former has a peppery odour and it is also present in nutmeg, and the latter has a warm woody odour.

**Grains of paradise** release 6-paradol, which has a spicy, peppery flavour.

© Shutterstock

Other popular brands of gin are Beefeater®, Gordon's®, Booth's®, Greenall's®, Seagram's®, Tanqueray®, Plymouth® and many  more.

Gin needs to be diluted with something if it is to be enjoyed, and the mixer of preference is usually tonic water, a drink that was invented in India. G&T is now drunk all around the world, and the tonic can either be of the sugar variety or the artificial sweetener kind. A 150 ml measure of the former will add 33 calories to the drink, whereas a 150 ml can of the latter will provide only 2 calories. Both rely on artificial sweeteners to counteract the bitter taste of the quinine. Quinine has several curative properties: it will reduce fevers, kill pain, and is antimalarial. However, the amount in tonic is small and a 150 ml can only contains around 10 mg. A typical dose of quinine for treating malaria is 50 times this amount, to be taken three times a day for seven days.

## Schweppes® Indian Tonic Water

This is the traditional type of tonic water.

*Ingredients*

| | |
|---|---|
| Carbonated water | Flavourings including quinine |
| Sugar | Sweetener (sodium saccharin) |
| Citric acid | |

**Carbonated water** is just $CO_2$ dissolved in water, which forms a little carbonic acid, which is a weak acid.

**Sugar** is sucrose.

**Citric acid** makes the drink slightly more acidic (pH around 3) – more so than the carbonic acid – and it has a citrus-like (lemon) flavour.

**Flavourings including quinine**. Whilst the flavourings are not identified, the quinine has to be, because some people are hypersensitive to it. Quinine has a very bitter flavour.

**Sweetener (sodium saccharin)** (food code E954) is an artificial sweetener, but with a somewhat bitter aftertaste.

## Schweppes® Slimline Tonic Water

This is the mixer you get when you ask for a gin and slim.

*Ingredients*

| | |
|---|---|
| Carbonated water | Flavourings (including |
| Citric acid | quinine) |
| Acidity regulator (E331) | Sweetener (aspartame) |

**Carbonated water**: see above.

**Citric acid**: see above.

**Acidity regulator** (food code **E331**) is sodium citrate, which – together with the citric acid – will buffer the drink and keep the pH at the desired level of around 3.

**Flavourings (including quinine)**: see above.

**Sweetener (aspartame)** is a more intense sweetener than saccharin, and in this case, it replaces all of the sugar in the original product.

Other makes of tonic water are available such as Bottle Green® and Fever-Tree®.

# 4. TOYS

**Play-Doh**®

*The Chemistry*

Dough can be made from various types of flour but this product needs to have a high level of gluten, which is what gives dough its elasticity, and wheat flour is the best in that respect. Gluten is 90% protein polymer, and this allows the dough used for baking bread to rise without losing the $CO_2$ gas that is being produced by yeast enzymes. In Play-Doh® the dough remains stable because other ingredients prevent the formation of this gas.

American wheat gives the best dough; its gluten content is high, its protein polymer chains are long and stable, and sulfur–sulfur bonds along the chains make them particularly flexible.

*Contents*

| | |
|---|---|
| Dough | Sodium borate |
| Mineral oil | Polyethylene glycol monostearate |
| Salt | Fragrance |
| Amylopectin | Dyes |
| Aluminium sulphate | |

**Dough** is made from American wheat flour that has a high gluten content.

**Mineral oil** is also known as liquid paraffin and it consists of long hydrocarbon chains. It acts as a lubricant for the molecules of the dough, making them easier to manipulate. It also coats these molecules, thereby making them less sticky. Unlike vegetable oils, mineral oils will not support the growth of microbes, so Play-Doh® does not go rancid over time.

**Salt** is sodium chloride and this ensures that the water in Play-Doh® cannot be used by microbes as a breeding ground. Here the salt content is high and ten times the amount that is added to dough used in baking.

**Amylopectin** is a natural long chain polymer consisting of linked glucose units with side chains attached, and it is better known as starch.

**Aluminium sulphate** attracts groups of gluten molecules, thereby making the dough slightly stiffer. It also has a bitter taste so that a child is unlikely to keep Play-Doh® in its mouth for long.

© Shutterstock

**Sodium borate** is a soft, natural mineral added to prevent microbes breeding in the dough, and if ants are attracted to it, they will be poisoned by it – see the section on ant killer in Chapter 12.

**Polyethylene glycol monostearate** is an emulsifier for oil and water mixtures, and it ensures that the mineral oil blends nicely into the dough.

**Fragrance** is now mainly vanilla although previously it was almond essence.

**Dyes** of different types are used to colour the Play-Doh®, but all are certified as safe.

# CHAPTER 11

# The Garage and the Car

© Shutterstock

You don't expect to read a list of ingredients on the products you use on your car or that you store in the garage. However, in some cases the information about their contents can be accessed *via* their material safety data sheets (MSDS) but they only reveal the ones that are potentially hazardous. Some companies are

Chemistry at Home: Exploring the Ingredients in Everyday Products
By John Emsley
© John Emsley, 2015
Published by the Royal Society of Chemistry, www.rsc.org

reluctant to say what else their product contains, claiming that this is commercially sensitive information.

In this chapter we will look at some products that are specifically for the car, some that are used in or outside the house, and some that are needed to clean up afterwards. They are the following:

1. Cleaning the car exterior (Turtle Wax® Car Wash)
2. Cleaning the car interior (Comma® Interior Cleaner, AutoGlym™ Leather Cleaner)
3. Windscreen wash (Xstream® Screenwash)
4. Hand cleanser (Swarfega®)
5. uPVC cleaner (Thompson's® uPVC Restorer)
6. Threadlocker (Loctite® Threadlocker Blue 242®)
7. Engine cleaner (Gunk®)
8. Drive cleaner (Thompson's® Oil and Drive Cleaner)

## 1.  CLEANING THE CAR EXTERIOR

*The Chemistry*

The dirt on the bodywork of your car will consist of the grime from the exhausts of other cars, dirt from the roads on which you drive, dust, and other pollutants, and possibly bird droppings and sap from trees. Some will stick to the metal and windows of your car, often quite tenaciously, because of the grease they contain. This can all be removed using a surfactant and you really need one that is strong.

There are ways to clean your car effortlessly if you pay a carwash or someone else to do it. Alternatively you can do it yourself at home. Simply adding washing-up liquid to the cleaning water will remove most of the dirt and grime but a more industrial type surfactant will guarantee you a better result.

**Turtle Wax® Car Wash**

Only certain ingredients have to be notified, namely the surfactants and antimicrobial agents. The makers do not reveal the ingredient that produces the 'hard shell shine', which may be a

© Shutterstock

silicone-based compound rather than a wax *per se*. However, some carnauba wax may be included, this being the wax that is used when you have your car waxed professionally.

*Contents*

Water
Potassium chloride
Sodium dodecylbenzenesulfonate

Sodium C14–16 olefin sulfonate
Coconut diethylamide

This product is basically a detergent mixture of proven surfactants that are not too harsh on the hands, even if you do not wear rubber gloves when washing the car.

**Water** is the solvent for the other ingredients, and these are present in amounts of less than 5% of the total.

**Potassium chloride** is simply there to keep all the ingredients well mixed and the liquid viscous.

**Sodium dodecylbenzenesulfonate** and **sodium C14–16 olefin sulfonate** are anionic surfactants and they produce lots of foam.

**Coconut diethylamide** makes the Turtle Wax® viscous and also helps produce foam.

There are more than 40 different kinds of car-wash products available. One made by Bilt-Hamber was judged the best in 2014 by *Auto Express* followed by Halfords and AutoGlym™.

## 2. CLEANING THE CAR INTERIOR

*The Chemistry*

Accidents will happen, especially if you have children in your car and they have drinks or snacks. Then it may be necessary to clean the seats. You could have it valeted, or you could do it yourself, and there are special cleaners that will help. Again these contain mainly surfactants and the other ingredients are likely to be a fragrance chemical and possibly a deodorant such as cyclodextrin.

### Comma® Interior Cleaner

This was the brand that *Auto Express* said in 2014 was the best at shifting the stains from upholstery.

*Contents*

| | |
|---|---|
| Water | Coconut diethanolamine |
| Monoethanolamine lauryl sulfate | Alcohol ethoxylate |

**Water** is the solvent.
**Monoethanolamine lauryl sulfate** is an anionic surfactant.

© Shutterstock

**Coconut diethanolamine** is a neutral surfactant that also ensures the cleaner is slightly alkaline.

**Alcohol ethoxylate** is a non-ionic surfactant.

Cleaning the interior can equally well be done with a household cleaner of the kind you would use on upholstery and carpets.

### AutoGlym™ Leather Cleaner

If you can afford real leather seats in your car then these will need to be treated with respect. The surfactants in this cleaner are more akin to those used indoors.

*Contents*

Water
Alkyl amido betaine
Alkyl polyglycosides

**Water** is the solvent.

**Alkyl amido betaine** is an amphoteric surfactant that gives lots of foam.

**Alkyl polyglycosides** are non-ionic surfactants that are gentle in action.

### 3.  WINDSCREEN WASH

*The Chemistry*

The windscreen of a car sometimes needs cleaning when it becomes splattered with road salt and dirt in winter and dead insects and dust in summer. In either situation, simply having water in your windscreen fluid is not very satisfactory, and indeed on a cold winter morning this may well be frozen solid. However, a couple of simple chemicals can solve the problem and these are ethanol, better known as alcohol, and ethane-1,2-diol, better known as ethylene glycol. Combinations of these will remain liquid down to the lowest temperatures.

### Xstream® Screenwash

The motoring magazine *Auto Express* reviewed the many screen-wash liquids on the market in 2013 and ranked this brand the most successful of 12 they tested. This is added to the screenwash water, and the amount added depends on the time of year.

In winter, the screenwash liquid should be 25% Xstream® to 75% water, or even a 50:50 mixture for extremely cold conditions when the temperature might fall to as low as −25 °C. In summer, the water in the screenwash tank need only have 5% of Xstream®, or 10% in autumn.

*Contents*

Ethanol
Ethylene glycol

**Ethanol** is simply alcohol and acts as a solvent. It remains liquid down to −114 °C and is the main constituent and accounts for as much as 90% of the total volume of this liquid.

**Ethylene glycol** is similar to ethanol, although as a solvent it is better able to loosen insect debris. As a pure liquid it freezes at −13 °C, but when used in conjunction with ethanol it can cope with temperatures as low as −26 °C.

Other brands of screenwash in the 2013 *Auto Express* tests were Prestone® BugWash®, which came second, and Halfords® Advanced Screenwash Double Concentrate, which came third.

## 4. HAND CLEANSER

*The Chemistry*

When you work on your car then your hands are likely to end up covered with dirty grease, which is not easily removed with just soap and water, or even washing-up liquid and water. What you need is something that can disperse the grease and which contains a mineral that can *ad*sorb it; in other words the grease will cling to the mineral particles and can be wiped away.

### Swarfega®

This product was designed to remove ingrained dirty grease and is mainly used in industry.

*Contents*

Aqua
Kaolin
Cetearyl alcohol
Dicaprylyl carbonate
Glycerin
Zanthan gum

Sodium lauryl sulfate
Phenoxyethanol
Diazolidinyl urea
CI 19585
CI 19140

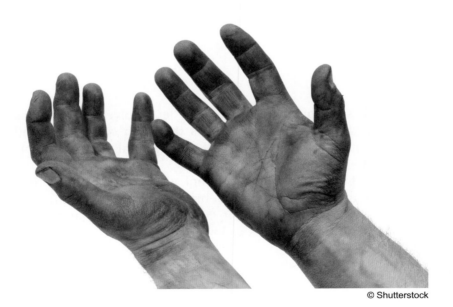

© Shutterstock

**Aqua** is water and a solvent.

**Kaolin** is also known as china clay and is the mineral kaolinite, which is aluminium silicate. This can adsorb other things very effectively, such as bacteria, organic molecules, pigments, dyes, and metals.

**Cetearyl alcohol** acts as a thickener to form the gel, as well as being a non-ionic surfactant.

**Dicaprylyl carbonate** is both an emollient and a skin conditioner, and is used in personal care products.

**Glycerin** stops the product from drying out.

**Zanthan gum** is a carbohydrate polymer that thickens and stabilises the gel.

**Sodium lauryl sulfate** is an anionic surfactant.

**Phenoxyethanol** is an antibacterial agent.

**Diazolidinyl urea** is a preservative used in skincare products and cosmetics.

**CI 19585** and **CI 19140** are yellow dyes.

## 5.  UPVC CLEANER

*The Chemistry*

Most window frames for double glazing are now made of the plastic uPVC. The u stands for unplasticised, meaning that no softening agent has been added to the polyvinyl chloride (PVC) polymer in order to make it flexible. (Most PVC we encounter has been treated to make it flexible, as with electric cables.)

Although uPVC can be made the colour of wood, much that is used for double glazing is white and will eventually become dirty and tired looking, but it can be cleaned. What is needed is a combination of abrasives, surfactants, solvents, and antimicrobial agents to remove fungi like *Aspergillus niger*, which is known as black mould. None of the ingredients must damage the uPVC, and ideally the cleaner should leave a protective film on the surface.

### Thompson's® uPVC Restorer

*Contents*

Aqua
Aluminium silicate hydroxide
Quartz/silica flour
Distillates (petroleum),
  hydrotreated light
Distillates (petroleum), hydrotreated
  heavy naphthenic
Solvent naphtha (petroleum), heavy
  aromatic
Triethanolamine

Amide, coco, *N,N*-bis(hydroxyethyl)
Aluminium oxide (non-fibrous)
Kerosine (petroleum), sweetened
DMDM hydantoin
Diethanolamine
Acrylic polymer
Acrylic polymer (cross-linked)
Non-ionic surfactant
Formaldehyde

*Abrasives*

These should be gentle abrasives, strong enough to dislodge the dirt but not strong enough to damage the uPVC. The ones in this product are **aluminium silicate hydroxide**, which is also known as china clay, **quartz/silica**, which is another mineral, finely ground so it has the consistency of flour, and **aluminium oxide (non-fibrous)**, which also acts as a mild abrasive.

*Surfactants*

**Amide, coco, *N,N*-bis(hydroxyethyl)** is a non-ionic surfactant. Also listed is '**non-ionic surfactant**', which is unspecified, probably because it comes as an impurity in one of the other chemical ingredients.

*Solvents*

**Aqua** is water and a solvent.

The non-aqueous solvents are **distillates (petroleum), hydro-treated light, distillates (petroleum), hydro-treated heavy naphthenic, solvent naphtha (petroleum), heavy aromatic** and **kerosine (petroleum), sweetened.** These are hydrocarbon solvents and as such will dissolve all kinds of grease. The last on this list, which is called paraffin in the UK, has been 'sweetened,' which means that any sulfur compounds that it might have contained have been removed. The others have been 'hydro-treated,' which means they have been reacted with hydrogen to remove unsaturated bonds.

**Triethanolamine** and **diethanolamine** are solvents that dissolve certain kinds of molecules that neither water nor hydrocarbon solvents will dissolve. They also have some surfactant ability.

*Antimicrobials*

**DMDM hydantoin** is an antimicrobial agent that works by releasing formaldehyde, which kills any fungus on the uPVC. (**Formaldehyde** is listed as an ingredient for this reason.)

*Protective agents*

**Acrylic polymer** and **acrylic polymer (cross-linked)** are sodium polyacrylate. These leave a film on the surface of the uPVC, which acts as the antistatic agent to prevent dust being attracted to the surface in future.

Other uPVC cleaner brands are Ronseal® and De-Sol-Vit®, and there are several others.

## 6. THREADLOCKER

*The Chemistry*

Threadlocking fluid is an adhesive applied to the threads of screws and bolts to prevent them from becoming loose due to vibrations and maybe even falling out. (Some threadlockers are even designed to act as gaskets and offer a permanent solution.)

What a threadlocker fluid needs is a polymer precursor and then something to make it polymerise so that it fills the gap with a solid material. This requires a catalyst to cause the reaction to occur and ideally this should start when the liquid is in place and not before. The catalyst itself is activated by the metal surface.

### Loctite® Threadlocker Blue 242®

This threadlocker will cope with most jobs, and the gap it has plugged is easily undone with ordinary tools.

*Contents*

Polyglycol dimethacrylate (60–100%)
Polyglycol oleate (10–30%)
Silica, amorphous, fumed, crystal-free (1–5%)

Cumeme hydroperoxide (1–5%)
Propane-1,2-diol (1–5%)
Cumene (0.1–1%)

**Polyglycol dimethacrylate** is the monomer from which the polymer forms and which then plugs the gap between the screw or bolt and the surface it is being attached to. The solution of this flows easily into the joint and there it begins to polymerise rapidly once it has been activated.

**Polyglycol oleate** acts as an emulsifier to keep the contents well mixed.

**Silica, amorphous, fumed, crystal-free** is micro-sized particles of silicon dioxide and it acts both as a filler and to provide extra strength.

**Cumeme hydroperoxide** is the oxidising agent that initiates the polymerization, and it does this by being activated by the metal surfaces that are being joined together. (**Cumene** is a trace impurity of the hydroperoxide.)

**Propane-1,2-diol** is a solvent for the ingredients.

Examples of other threadlocker brands are Scotch Weld™ (from 3M®) and Permatex®.

## 7. ENGINE CLEANER

*The Chemistry*

People may need to look under the bonnet of their car to check oil and water levels. However, there are those who like to do more and, if they have an older car, they may then find that a lot of grease and dirt has accumulated. If so, there are products to help remove it. (Nowadays, emission regulations mean that modern engines are much cleaner.)

What is needed is something to soften and remove the grease because some of this will have become oxidised and hardened. There also needs to be something to deal with the dirt that comes from road surfaces and other cars' exhausts.

### Gunk®

Although this comes in several forms, the spray version is the one that most people would use because it can hit those parts of the engine that are difficult to reach. You spray it on the grime, leave for 15 minutes, and then wash off with water.

*Contents*

| | |
|---|---|
| Propane | Ammonia |
| Butane/isobutane | Aromatic hydrocarbon |
| Sodium tripolyphosphate | 2-Butoxyethanol |
| Triethanolamine (90%) | |

- **Propane, butane,** and **isobutane** are the propellant gases. These are under pressure in the canister and they force out the content through a nozzle when the pressure is released.
- **Sodium tripolyphosphate** is best known for being a water softener, but it can bond to most metals and it will lift off any rust.
- **Triethanolamine** is used to solubilise the sticky gunge.
- **Ammonia** attacks grease and helps to solubilise it.
- **Aromatic hydrocarbon** is an excellent grease dissolver and it consists of various benzene-based chemicals (although not benzene itself, which is carcinogenic).

**2-Butoxyethanol** is a solvent that will happily dissolve both hydrocarbons and water-soluble grime. It also has some surfactant properties.

Other engine cleaners include AutoGlym™ and own-brand products.

## 8.   DRIVE CLEANER

*The Chemistry*

Sometimes there will be unsightly oil stains on the floor of the garage, on the drive leading to the garage, or on the off-road parking area in front of the house. These can be difficult to remove if they have seeped into the surface of concrete or paving stones.

Removing them is not a problem if you use a combination of powerful surfactants together with sodium hydroxide, which has the power to chemically attack grease and solubilise it. Alternatively it may be possible to use a pressure hose to ensure complete removal after treatment.

### Thompson's® Oil and Drive Cleaner

This product is capable of removing stubborn stains of long-standing if you treat them with it, leave for about half an hour, then brush them with a stiff brush, and finally wash them away. However, a second application may be needed.

*Contents*

Aqua
Sodium dodecylbenzene sulphonate
2-2'-2''-Nitrotriethanol
Amide, coco, *N,N*-bis(hydroxyethyl)

Sodium alkylether sulfate
Diethanolamine
Sodium hydroxide

**Aqua** is water and a solvent.

**Sodium dodecylbenzene sulphonate** is a powerful anionic surfactant that foams easily.

**2-2'-2''-nitrotriethanol** is better known as triethanolamine and is good at solubilising grease, to the extent that it can be washed away with water.

**Amide, coco, *N,N*-bis(hydroxyethyl)** is a non-ionic surfactant.

**Sodium alkylether sulfate** is an anionic surfactant.

**Diethanolamine** has some surfactant properties and it can be a solvent for acidic residues.

**Sodium hydroxide** makes the product alkaline, which helps with grease removal.

Drive cleaning products are also sold under the names of Gunk®, Swarfega®, Stone Care™, and others.

CHAPTER 12

# The Garden Shed

© Shutterstock

Many homes have a garden, and many gardens have a shed. This might well be somewhere for a man to retreat to, as advocated in Gordon Thornburn's 2002 book *Men and Sheds*, but for most of us it's where we keep the things we use in or around the garden. If we are a keen gardener, then it is the place where we keep

Chemistry at Home: Exploring the Ingredients in Everyday Products
By John Emsley
© John Emsley, 2015
Published by the Royal Society of Chemistry, www.rsc.org

fertilizers and pesticides, but it can also be a place to store other products that we may need when we relax outside in summer.

There are hundreds of products designed to assist the gardener and those discussed here are only a few of them, but I have tried to cover all categories, including a few other items that might usually be kept in the shed. The products in this chapter are:

1. Fertilizer (Growmore®)
2. Weedkiller (Roundup® and Verdone® Extra Lawn Weedkiller)
3. Slug pellets (Eraza® Slug & Snail Killer)
4. Insecticide (Provado® Ultimate Bug Killer)
5. Insect repellent (Jungle Formula®)
6. Ant killer (Nippon® Ant Killer Liquid, Nippon® Ant Killer Powder, and Nippon® Ant Killer Control System)
7. Disinfectant (Jeyes® Fluid)
8. Wood preservative (Sadolin® Classic)
9. Decking cleaner (Ronseal® Decking Cleaner)
10. Garden furniture cleaner (Ronseal® Garden Furniture Wipes)
11. Fireworks (Roman Candle)
12. Outdoor lozenge (Fisherman's Friend® Cherry Flavour)

## 1.  FERTILIZER

*The Chemistry*

Fertilizers are needed to replace the elements that are extracted by plants from the soil, if it is not to become depleted. There are two kinds of fertilizer: the organic kind – namely animal manure, compost, and bone and fish meal – and the inorganic kind. The latter are referred to in terms of their element names: nitrogen, phosphorus and potassium (N-P-K). It is best if some of these are available at the start of the growing season, while some might be released to the soil more slowly through the growing season.

Plants also need other elements that are essential for their growth, namely calcium, sulfur and magnesium, but these are not needed to the same extent and the soil can usually provide all that is required, although adding more will do no harm and may improve plant quality. Finally there are the micronutrients, elements that are also needed by plants but only in tiny amounts. These are boron, iron, manganese, zinc, copper, and molybdenum.

© Shutterstock

## Growmore®

This is a 7-7-7 mixture of N-P-K meaning it has equal proportions of the three major nutrients. Nitrogen is needed for growth, phosphorus for flowers and fruit, and potassium for roots. The 7-7-7 is an old way of showing the ratio of these elements as their oxides, which is not the form in which they are present in the fertilizer itself. In actual percentages of the elements, the amounts are in the ratio of 7-3-6. This fertilizer also provides trace metals.

*Contents*

Ammonium sulphate
Urea
Mono and di-ammonium phosphate
Superphosphate
Triple superphosphate
Rock phosphate
Potassium chloride (muriate of potash)
Potassium sulphate

Calcium nitrate
Urea formaldehyde
Ferrous sulphate
Inert fillers such as sand or limestone
Coating materials such as oil, clay or talc
Secondary nutrients and/or micronutrients

**Ammonium sulphate** supplies both nitrogen and sulfur in a soluble form.

**Urea** is a source of nitrogen.

**Mono and di-ammonium phosphate** supply nitrogen and phosphorus, which are instantly available.

**Superphosphate** and **triple superphosphate** are calcium dihydrogen phosphate, which is phosphate rock made soluble by treatment with acid, and these are soluble in soil water.

**Rock phosphate** is calcium phosphate in the mineral form and is a slow-release source of phosphate.

**Potassium chloride**, also known as muriate of potash, is a slow-release source of potassium.

**Potassium sulphate** is a readily available source of potassium.

**Calcium nitrate,** also known as nitro chalk, provides calcium as well as nitrogen. Plant roots can absorb nitrate.

**Urea formaldehyde** is a slow-release source of nitrogen.

**Ferrous sulphate** provides both iron and sulfur.

**Inert fillers such as sand or limestone;** these are silicon dioxide and calcium carbonate.

**Coating materials such as oil, clay or talc** stop the fertilizer pellets from sticking together.

**Secondary nutrients (micronutrients)** are trace amounts of the elements zinc, magnesium, manganese, copper, zinc, molybdenum, and boron.

## 2.  WEEDKILLER

*The Chemistry*

All kinds of things will kill unwanted plants, and a hundred years ago things like arsenic were used. Today we have weedkillers that work more effectively; some are even selective weedkillers and are able to kill unwanted plants that grow on the lawn while leaving grass unaffected.

What we demand of a garden weedkiller is that it works well but presents no threat to children, animals, or pollinating insects. One of the most popular weedkillers is Roundup®, which is based on glyphosate. This relatively simply chemical inhibits the enzyme that plants need in order to make the essential amino acids phenylalanine, tryptophan, and tyrosine. Glyphosate is an amino acid derivative that blocks this part of their metabolism.

### Roundup®

*Contents*

| | |
|---|---|
| Water | Isopropylamine |
| Isopropylammonium glyphosate | Polyoxyethylene alkylamine |
| Ethoxylated tallow amine | FD&C Blue no.1 |

**Water** is the solvent.

**Isopropylammonium glyphosate** is the active agent.

**Ethoxylated tallow amine** is a surfactant that wets the surface of the weed thereby making it easier for the glyphosate to penetrate.

**Isopropylamine** ensures that all the glyphosate is in the active form of the negative glyphosate ion.

**Polyoxyethylene alkylamine** is an emulsifier, and it also wets the leaves and ensures the active agent comes in contact with them.

**FD&C Blue no.1** is a colourant.

Other weedkillers include Resolva®, whose active ingredient is butafenacil, and Pathclear®, which contains glyphosate, butafenacil, and oxadiazon.

### Verdone® Extra Lawn Weedkiller

This is a selective weedkiller, designed to kill weeds but leave grass unaffected. It contains three kinds of weedkiller targeted at various weeds such as dandelions, daisies, clover, ragwort, and moss. The product is diluted with water and applied with a watering can.

*Contents*

MCPA (ISO) >25%
Fluroxypyr 10–25%
Clopyralid 5–10%

**MCPA (ISO)** is short for 2-methyl-4-chlorophenoxyacetic acid iso-octyl ester and it works against broadleaf weeds like clover, dandelion, ragwort, and dock.

**Fluroxypyr** is a systemic broadleaf herbicide and kills ragwort and dock.

**Clopyralid** is effective against thistles and clover.

Other selective weedkillers are Weedol® and Yates® Weed'n'Feed™.

## 3.  SLUG PELLETS

*The Chemistry*

Slugs will eat almost anything in the garden, and so will snails. Killing them should only be done with a poison that does not threaten other forms of wildlife such as birds and hedgehogs, and the poison must not attract young children. They can be dissuaded from eating slug pellets by incorporating a bittering agent. The poison used in slug pellets is metaldehyde, which these pests cannot sense and it kills them fairly quickly. It is mixed with bait such as bran or wheat.

### Eraza® Slug & Snail Killer

This product comes as tiny pellets protected by a polymer layer, thereby making it more effective and weather resistant. In this form there can be less of the poison per pellet and yet it is still effective. The pellets are known technically as TR3799 and produced by Chiltern Farm Chemicals Ltd, which produces slug pellets mainly to protect farm crops.

*Contents*

| | |
|---|---|
| Bran or wheat | Polymer |
| Metaldehyde | Blue dye |
| Bitrex | |

**Bran or wheat** is the bait.

**Metaldehyde** is the poison and pellets contain 1.5% of this material. There is enough in one pellet to kill a slug or snail.

**Bitrex** is denatonium benzoate, and it must be added by law to slug pellets to deter other creatures from eating them. It is the bitterest substance known, but slugs cannot taste it so they continue eating the bait.

**Polymer** is not specified but is designed to keep the pellets from decomposing when it rains, and it has to be edible to slugs so is probably a natural polymer based on cellulose.

**Blue dye** is a warning to humans that these pellets are not edible.

Other brands of slug pellet are Bayer®, Doff®, Slug Gone® *etc.*, and all use very similar chemicals to Eraza®.

## 4.   INSECTICIDES

*The Chemistry*

The garden can be infested with insects that attack flowers, fruit, and vegetables. Greenfly, blackfly, whitefly, thrips, and others can be kept at bay for several weeks with a systemic insecticide that pervades the plant and poisons any insect that tries to eat its leaves or suck its sap. Pollinating insects like bees are not affected.

### Provado® Ultimate Bug Killer

This product needs to be diluted before use although there are versions that are already made up in spray-type containers.

*Contents*

Thiacloprid
1,2-Propanediol

- **Thiacloprid** is the active agent and is present at a concentration of 15 mg per litre. The concentrated form of this insecticide contains 9 grams per litre.
- **1,2-Propanediol** is the solvent for the insecticide, which is then diluted with water before use.

Other insecticides are available, such as Trounce®, and these may be based on natural chemicals like pyrethrum and, like the thiacloprid in Provado, are approved by the Royal Horticultural Society.

## 5.   INSECT REPELLENT

*The Chemistry*

At certain times of year we might expose ourselves to insect attack, perhaps on holiday, on a walk in the countryside, while picnicking, or while gardening. Certain locations and times are particularly prone to insects. For example, the highlands of Scotland are notorious for midges from June to September. The answer is to apply an insect repellent and the most effective ones contain diethyl toluamide (also known as DEET). When this is applied to exposed skin, the DEET is slowly released into the air around you. This chemical confuses an insect's ability to detect you as a source of animal blood and so deflects it from attacking you.

# Jungle Formula®

*Contents*

| | |
|---|---|
| Diethyl toluamide (DEET 50%) | Aqua |
| Alcohol denat. | Dipropylene glycol |
| Butane | Panthenol |
| Isobutane | Parfum |
| Propane | |

© Shutterstock

**Diethyl toluamide (DEET 50%)** is the chemical that repels the insects.

**Alcohol denat.** is ethanol and a solvent. It has had a bittering agent added so that it cannot be misused.

**Butane, isobutene,** and **propane** are the gases that provide the pressure for the aerosol.

**Aqua** is water and a solvent.

**Dipropylene glycol** is a solvent.

**Panthenol** is related to vitamin $B_5$ and is thought to have beneficial effects on the skin.

**Parfum** is not specified, which means it does not contain any of the fragrance ingredients to which some people might be sensitive.

Other DEET-containing insect repellents are Care Plus®, Off!®, En Garde!® *etc.*

## 6.   ANT KILLERS

*The Chemistry*

Ants can be a problem, both in the garden and in the house if they can find a way in and locate food. What is needed is a bait containing boric acid, which they will take back to the nest and which takes time to act. They pass the message of food availability to other ants in the nest who then also consume it.

An alternative ant killer is to sprinkle a contact poison around the nest entrances and along runs which the ants are using.

## Nipon® Ant Killer Liquid

*Contents*

Sugary liquid
Boric acid 3.2%

**Sugary liquid** is a blend of corn syrup and a small amount of honey.

**Boric acid** is the toxic agent that affects an ant's digestive system. Other insects like cockroaches are also susceptible to it, yet as far as humans are concerned it is no more toxic than common salt.

## Nippon® Ant Killer Powder

*Contents*

Permethrin 0.488%
Talc base

**Permethrin** is the insecticide. It is safe to use in and around the home if properly applied.

**Talc base** is magnesium silicate and it is a natural white mineral renowned for its softness.

This product can be dusted into nooks and crannies, around doors, windows, air bricks, and other places where ants have been observed coming into or moving around the home. It can also be applied to the entrances of ants nest outdoors.

## Nippon® Ant Killer Control System

This relies on the insecticide spinosad and this is taken back to the colony and even fed to the queen ant. It is particularly effective against the common black ant.

*Contents*

Spinosad 0.081%
Sugary liquid

**Spinosad** is the insecticide, based on a natural agent. It has to be placed under a cover to prevent other insects like bees from coming in contact with it.

**Sugary liquid** is a blend of corn syrup and a small amount of honey.

Other brands of ant killer are available, such as Raid®, Ant Stop!™ and Terro®.

## 7.  DISINFECTANT

*The Chemistry*

There are times when garden equipment, greenhouses, seed trays, pet areas, outdoor surfaces, and wheelie bins need to be disinfected. This can be done using bleach but there is a traditional chemical, Jeyes® Fluid, which was first produced in the 1870s and has been used in the gardens of the Royal estates in the UK since 1896. It was originally formulated from the by-products of the coal tar produced by gasworks, of which there was one in every town and city in the UK before the advent of natural gas. Jeyes® Fluid relies on the effectiveness of various antimicrobial agents.

### Jeyes® Fluid

*Contents*

| | |
|---|---|
| Aqua | PEG |
| Isopropyl alcohol | 4-Chloro-*m*-cresol |
| Sodium castorate | Tar acids |
| Terpineol | |

**Aqua** is water and a solvent.

**Isopropyl alcohol,** also known as propan-2-ol, is a solvent.

**Sodium castorate** is the sodium salt of castor oil which is chiefly composed of ricinoleic acid, a monounsaturated 18-carbon chain fatty acid

**Terpineol** is also known as terpene alcohol. It is a component of tall oil, which is extracted from pine trees and it is mainly oleic acid based. It will dissolve grease and it has a pine-like odour.

**PEG is polyethylene glycol** which is used to stabilise the emulsion.

**4-Chloro-*m*-cresol** is also known as PCMC and has antiseptic and fungicidal properties.

**Tar acids** also have disinfectant properties and are produced by heating materials such as wood to temperatures of around 700 °C and collecting the liquid which distils. They consist of phenol with various hydrocarbon chains attached.

## 8.  WOOD PRESERVATIVE

*The Chemistry*

Nature has evolved wood to endure for hundreds of years and as such it is protected by the bark of the living tree. However, when the tree is cut down and the wood put to use, then it may be exposed to the weather, to oxidation, to UV, and to microbial attack. What is needed is a paint or varnish that will penetrate the wood and coat the surface with a film of polymer that contains biocides.

## Sadolin® Classic

This product is suitable for all kinds of exterior woodwork around the garden, such as fences, sheds, and decking. It describes itself as alkyd resin dissolved in dearomatised white spirit, which means it is virtually free of benzene, a known carcinogen.

*Contents*

Alkyd resin
4,5-Dicholor-2-octyl-2H-isothiazol-3-one
 (<1%)
Ethyl methyl ketoxime (<1%)
Naphtha (petroleum) hydro-treated –
 D40 dearomatised (30–60%)
Naphtha (petroleum) hydro-treated –
 D60 high flash (<1%)
Xylene (<1%)

2-Butoxyethanol (<1%)
2-Methylpentane-2,4-diol (<1%)
3-Iodo-2-propylbutyl carbamate
 (<1%)
Cobalt carboxylate (<1%)
Stoddard solvent, low bp naptha
 <1% benzene (<1%)
Tolylfluanid (<1%)

*Preservatives*

**Alkyd resin** provides a strong protective film on the surface of the wood when it dries.

**4,5-Dicholor-2-octyl-2H-isothiazol-3-one** is a biocide to protect the wood against microbial attack.

**3-Iodo-2-propylbutyl carbamate** is a biocidal agent.

**Tolylfluanid** is a fungicide.

*Solvents*

**Naphtha (petroleum) hydro-treated – D40 dearomatised** and **naphtha (petroleum) hydro-treated – D60 high flash** are hydrocarbon solvents (long chains ones and cyclic ones) with a boiling range of 155 to 217 °C, which means they evaporate slowly and are not likely to be easily ignited. (The flash point of a solvent is the lowest temperature at which its vapour can be ignited; the higher the flash point the safer the solvent.)

**Xylene** is a benzene ring with two methyl groups attached.

**2-Butoxyethanol** acts as a solvent for some of the ingredients as does **2-methylpentane-2,4-diol**.

**Stoddard solvent, low bp naptha <1% benzene** is another name for white spirit.

*Other Ingredients*

**Ethyl methyl ketoxime** prevents a skin forming on the surface of the product before it is used.

**Cobalt carboxylate** promotes the curing of the alkyd resin so it delivers a permanently strong surface layer.

Other wood preservatives are marketed under the brand names Cuprinol® and Ronseal®, the latter being advertised with a phrase that has now entered the English language: 'it does exactly what it says on the tin®'.

© Shutterstock

## 9.   DECKING CLEANER

*The Chemistry*

Over the space of a year any decking in a garden will tend to accumulate an unsightly film of algae, moss, and dirt that needs cleaning off. As with most cleaning, this requires a combination of surfactant, water softener, and something that will leave behind a film with biocidal activity.

## Ronseal® Decking Cleaner

*Contents*

Aqua
Potassium tripolyphosphate
Alcohol ethoxylate

2,2',2''-(Hexahydro-1,3,5-triazine-1,3,5-triyl)triethanol
1,2-Benzisothiazol-3(2*H*)-one
Sodium hydroxide

**Aqua** is water and a solvent.

**Potassium tripolyphosphate** is a water softener that improves the action of the surfactant and does not form insoluble deposits.

**Alcohol ethoxylate** is a non-ionic surfactant.

**2,2',2''-(Hexahydro-1,3,5-triazine-1,3,5-triyl)triethanol** acts as a biocide and disinfectant.

**1,2-Benzisothiazol-3(2*H*)-one** is a powerful fungicide.

**Sodium hydroxide** keeps the solution alkaline, which helps disperse grease.

## 10.  GARDEN FURNITURE CLEANER

*The Chemistry*

Over winter, garden furniture left outdoors becomes covered with a layer of grime and small microbes such as mould and mildew, both of which are types of fungus. When spring arrives, it needs cleaning, polishing, and protecting, and this requires surfactants to clean it, a chemical to revive it, and something to kill the fungus. The cleaning can be undertaken with the usual products and water, or we can use a convenient wipe that contains all of them, and indeed can be used whenever outdoor furniture needs it.

### Ronseal® Garden Furniture Wipes

*Contents*

Aqua
Alcohols C9–11 ethoxylated
4-Isopropenyl-1-methylcyclohexene
  (D-limonene)
Rosin

DMDM hydantoin
Carbamic acid butyl-3-iodo-2-
  propynyl ester
Tetrasodium EDTA

**Aqua** is water and a solvent.

**Alcohols C9–11 ethoxylated** is a non-ionic surfactant.

**4-Isopropenyl-1-methylcyclohexene** is the chemical name for limonene, which is the fragrance of oranges.

**Rosin** is the pitch-like material obtained from pine trees. Its main component is abetic acid. It is a traditional oil used to restore wood that has become dry.

**DMDM hydantoin** (the DMDM stands for dimethylol dimethyl) is a powerful antimicrobial agent and works by releasing formaldehyde which kills the mould.

**Carbamic acid butyl-3-iodo-2-propynyl ester** kills mould such as mildew.

**Tetrasodium EDTA** counteracts hard water by sequestering calcium ions that would otherwise interfere with the surfactant.

## 11.   FIREWORKS

*The Chemistry*

Sometimes we want to celebrate an event by letting off fireworks and these are often best stored in the garden shed. Basically, a firework is something that burns dramatically and even with an explosive force to propel its contents into the sky where it may perform amazing displays. What a firework needs to do is burn with a temperature high enough to cause certain metals to display different colours, which they can do by having their electrons excited. As these electrons return to their normal state they give off visible coloured light, depending on the metal.

There are many kinds of firework and a popular one is the Roman candle, which expels coloured stars.

© Shutterstock

## Roman Candle

Within its thick-walled cardboard tube this consists of the following components in the order in which they ignite.

*Components*

| | |
|---|---|
| Blue touch paper | Delay composition |
| Delay composition | Green star |
| Red star | Gunpowder |
| Gunpowder | Clay or plastic base |

**Blue touch paper** is paper treated with potassium nitrate. When lit, a smouldering chemical reaction begins between the cellulose of the paper and the nitrate, and this eventually ignites the gunpowder at the top of the firework.

**Delay composition** has 62% potassium nitrate, 18% sulfur, and 20% carbon and burns to produces a small display of sparks, eventually igniting a small charge of gunpowder below the first star, which is then ejected.

**Gunpowder** is a mixture of potassium nitrate, sulfur, and charcoal in the ratio $1:1:6$.

**Red star** is composed of potassium chlorate, strontium nitrate, strontium carbonate, and aluminium powder, held together by a natural gum. The potassium chlorate provides the oxygen, which allows the aluminium to burn with an intense hot flame that stimulates the strontium ions to emit red light.

**Green star** consists of barium chlorate, potassium chlorate, and charcoal, held together by a mixture of tree resin and dextrin. Now it is the turn of the barium ions to emit light and this is green.

## 12.   OUTDOOR LOZENGE

*The Chemistry*

If you need to work in the garden, then you need something to keep your airways open whatever the weather. The traditional sweet to suck was one that was originally designed in the 1860s for those working on trawlers in the extreme environment of the North Sea.

## Fisherman's Friend® (Cherry Flavour)

*Ingredients*

| | |
|---|---|
| Sorbitol | Elderberry juice |
| Sucralose | Herbal flavour |
| Acesulfame K | Menthol |
| Cherry flavour | Magnesium stearate |

**Sorbitol** is a natural sweetener.

**Sucralose** is an artificial sweeter and so is **acesulfame K**, which is particularly used for chewable tablets because it can mask bitter tasting components.

**Cherry flavour** comes mainly from benzaldehyde.

**Elderberry juice** has been used in traditional medicine for hundreds of years and is thought by some to stimulate the immune system.

**Herbal flavour** is not specified.

**Menthol** occurs naturally in peppermint oil in the form of the left-hand molecule, and it can be manufactured. It appears to have a refreshing effect because it can trigger the cold-sensitive receptors when it comes in contact with the skin, the nose, or the throat, producing a sensation of freshness.

**Magnesium stearate** is both an excipient and a processing chemical. As the former, it helps the tablet to release its ingredients when moistened by saliva; as the latter, it prevents the tablets from sticking to the machinery during manufacture.

# Glossary

Chemicals are listed in alphabetical order under the names used in the various chapters. Here is extra information, consisting of the official name used by chemists, its molecular structure, and the use to which it is put in everyday products.

Sadly there is no agreed system for naming the kinds of chemicals used as ingredients and the same chemical may sometimes have several names. While most people now know that alcohol is ethyl alcohol, they might not realise that ethanol is its chemical name. Glycerine, which is also spelled glycerin, is also called glycerol, propanetriol, and 1,2,3-hydroxypropane – the last of these being the chemically preferred name.

*Acacia senegal* **gum** consists mainly of a long chain hydrocarbon called hentriacontane, which has 31 carbon atoms. It is used as a thickener.

**Acesulfame K** is potassium 6-methyl-2,2-dioxo-oxathiazine-4-one 2,2-dioxide and is an intense artificial sweetener. It has the food code E950.

**Acetic acid** is ethanoic acid, formula $CH_3CO_2H$, and is the chief constituent of vinegar. It has the food code E260. It is

Chemistry at Home: Exploring the Ingredients in Everyday Products
By John Emsley
© John Emsley, 2015
Published by the Royal Society of Chemistry, www.rsc.org

produced by anaerobic fermentation of carbohydrates or made industrially from methanol, which in turn is made from the methane of natural gas.

**Acrylates** are polymers of acrylic acid whose chemical formula is $H_2C=CHCO_2H$. When the acidic hydrogen **H** is replaced by a methyl group ($CH_3$) then we have methyl acrylate, and similarly there can be ethyl acrylate and butyl acrylate. When a mixture of these is polymerised we have **acrylate copolymer**, which can be a viscous liquid used as a thickener, an emulsifier, and as an antistatic agent in personal care products.

**Acrylate/steareth-20 methacrylate crosspolymer** – see acrylates and steareth. This is used as a thickener for water-based solutions so they become more viscous.

**Adipic acid** is hexanedioic acid, formula $HO_2C(CH_2)_4CO_2H$, with two acid groups, and has a pleasant fruity flavour. It has the food code E355. It can be used as a gelling additive.

**Alcohol ethoxylates (AE)** consist of a long chain alcohol ROH to which are attached ethoxy groups ($-CH_2CH_2O-)_n$, where $n$ may be 3–14. The number may be indicated in the name, such as C12–C14 AE-3. The formula for this would be $CH_3(CH_2)_{10-12}CH_2(OCH_2CH_2)_3OH$. AEs are also called polyoxyethylene alcohols (POE). They are used as non-ionic surfactants.

**Alkane/cycloalkane/aromatic hydrocarbons** is a mixture of solvents that will dissolve oils and oil-based materials. The alkane hydrocarbon may be indicated as, for example, C10–C14, meaning it consists of between 10 and 14 carbons in the chain. Cycloalkane means the hydrocarbons may be present as rings, such as cyclohexane with six carbons. Aromatic hydrocarbons refer to those with a benzene ring as part of their molecular structure, although benzene itself is not present as this is a cancer-causing chemical.

**Alkyd resin** is a polymer formed by the reaction of a long chain acid with an alcohol, both of which must have two or more of the necessary reacting groups to enable polymer formation to occur and allow for cross-linking by reaction with other ingredients.

**Alkyl dimethyl amine *N*-oxide** is $RN(O)(CH_3)_2$ where R consists of a hydrocarbon chain 10 to 16 carbons long. These are used as non-ionic surfactants.

**Alkyl polyglycosides** consist of a long chain fatty acid bonded to a chain of carbohydrates such as those of maize starch, and these are the glycoside component. When the carbohydrate is glucose then they are known as alkyl polyglucosides.

**Allantoin** is (2,5-dioxo-4-imidazolidinyl)urea. It is used as a moisturiser in personal care products.

**Aluminium hydroxide** is $Al(OH)_3$.

**Aluminium starch octenylsuccinate** is the aluminium salt of the reaction product of octenylsuccinic anhydride and starch. It is used as an anti-caking agent and to increase the viscosity of a liquid.

**Americium** is a man-made element, atomic number 95, and all its isotopes are radioactive. Americium-241 has a half-life of 432 years, decaying by the emission of alpha particles. It is used in smoke detectors.

**Ammonia caramel** – see caramel.

**Ammonium dihydrogen phosphate** is $(NH_4)H_2PO_4$, and **diammonium phosphate** is $(NH_4)_2HPO_4$. Both are used as fertilizers.

**Ammonium sulfate** is $(NH_4)_2SO_4$ and is used as a fertilizer.

**Antimicrobial agents** – see Technical Words.

**Antioxidants** – see Technical Words.

**Anthrocyanins** are natural pigments which can be red, blue, or purple, and they give colour to fruits like red grapes and blackcurrants. Their food code is E163.

**Arachidyl behenate** is a long chain fatty ester (cosyl, C20) of a long chain fatty acid (docosyl, $C_{22}$). It is used as a skin moisturiser and viscosity regulator.

**Aspartame** is an artificial intense sweetener consisting of two amino acids, aspartic acid bonded to phenylalanine, with a methyl group attached to the latter as its methyl ester. It is *N*-(L-α-aspartyl)-L-phenylalanine 1-methyl ester and has the food code E951.

**Aqua** is water, $H_2O$.

**Barium chlorate** is $Ba(ClO_3)_2$.

**Beeswax** consists of a saturated long chain fatty acid (palmitic, C16) bonded to a saturated fatty alcohol (oleic, C18). It is used as a glazing agent for foods and has the food code E901.

**Behentrimonium chloride** is $CH_3(CH_2)_{21}N(CH_3)_3^+$ $Cl^-$. It is attracted to strands of hair by virtue of its positive charge and acts as an antistatic agent.

**Benzaldehyde** is $C_6H_5CHO$ and is part of the natural flavour of almonds and cherries.

**Benzene C10–13 alkyl derivatives** consist of a hydrocarbon chain with a benzene ring attached at various possible positions along the chain. These are liquids with boiling points in the range 250 °C to 300 °C, depending on the blend of molecules. They are used to make linear alkyl benzene sulfonate surfactants.

**Benzisothiazolinone** is 1,2-benzisothiazol-3-one and is used as a powerful germicide. See also *N*-butyl-benzisothiazolinone.

**Benzophenone-n** refers to a group of molecules in which there are hydroxyl (OH) groups attached to the benzene rings of benzophenone. For example benzophenone-1 has two OHs on the same ring, while benzophenone-7 has one OH and one chlorine atom attached to the same ring, and benzophenone-10 has one OH and one methoxy ($CH_3O$) group on one ring and a methyl group ($CH_3$) on the other ring. They are used to protect against UV rays.

**Benzyl alcohol** acts partly as a solvent for other ingredients but also prevents the growth of bacteria.

**Benzyl salicylate** is $C_6H_5CO_2C_6H_4OH$ and is used as a fixative for fragrance molecules so they blend with the other ingredients. Of itself, it has almost no odour.

**Betaine** is trimethylglycine, $(CH_3)_3N^+CH_2CO_2^-$. When one of the methyl ($CH_3$) groups is replaced by a long chain hydrocarbon then the resulting molecule is called an **alkyl betaine**. When the long chain hydrocarbon contains an amido group (–CO–NH–) in its chain then the product is known as an **amido betaine**. Both kinds of betaine make good (neutral) surfactants and produce lots of foam.

**BHA** is short for butylated hydroxyanisole. It is an antioxidant and added to edible fats and oils. It has the food code E320.

**BHT** is short for butylated hydroxytoluene. It is a powerful antioxidant and has the food code E321. It is used in personal care products and foods.

**Bisabolol** is a natural complex alcohol believed to have skin-healing properties. It is extracted from the camomile plant.

**Boric acid** is $H_3BO_3$ and it is a mild antiseptic. It is used to preserve caviar (food code E284), to control nuclear reactors, and to kill ants.

**Buffering agent** – see Technical Words.

**Butane** and **isobutane** are gases and have the same formula, $C_4H_{10}$, but differ in their molecular structure.

**2-Butoxyethanol** is $CH_3(CH_2)_3O(CH_2)_2OH$ and it is an excellent solvent for both hydrophobic and hydrophilic compounds.

**N-Butyl-1,2-benzisothiazolin-3-one** is a pesticide active against insects, rodents, and fungi and is used to protect polymers such as silicones and PVC.

**Butyl acetate** is $CH_3C(O)OC(CH_2)_3CH_3$ and is a solvent with a boiling point of 127 °C.

**Butylene glycol** is 1,4-butanediol and is a humectant that is naturally present in skin.

**Butylhydroxyanisole** – see BHA.

**Butylhydroxytoluene** – see BHT.

**Butylphenyl methylpropional** is a fragrance molecule with an open-air freshness smell.

**C11–15 pareth-9, C14–15 pareth-7** – see pareth.

**C12–C18 alkyldimethyl amine *N*-oxide** is a powerful non-ionic surfactant. Its nitrogen atom is bonded to a hydrocarbon chain with between 12 and 18 carbon atoms.

**Caffeine** is 1,3,7-trimethylpurine-2,6-dione. It has three effects: it stimulates the brain by increasing the level of dopamine; it relaxes the airways so making breathing easier; and it releases energy from stores within the body.

**Calcium carbonate**, formula $CaCO_3$, occurs naturally as limestone, marble, and egg shells. It has the food code E170. Ground-up calcium carbonate makes an excellent scouring powder which does not scratch surfaces.

**Calcium hydroxide** is also known as slaked lime and is $Ca(OH)_2$. It is added to control pH and prevent acidity. It has the food code E527.

**Calcium nitrate** is $Ca(NO_3)_2$, and is also known as nitro chalk. It is used as a fertilizer.

*Calendula officinalis* is extracted from marigolds. It has anti-inflammatory and antioxidant properties. It contains triterpenoid esters, which have steroid-like molecular structures, and flavoxanthin. It can be used as an orange food colourant (food code E161).

**Camphor** is 1,7,7-trimethylbicycloheptanone. Although it occurs naturally, it is mainly produced by the chemical industry. It is used in body rubs and has a cooling sensation on the skin.

**Candelilla wax** has chains of up to 33 carbons and is extracted from the candelilla plant (*Pedilanthus macrocarpus*), which grows in Mexico. It melts at 67 °C and is used in place of beeswax.

**Caprilic/capric glycerides** are saturated oils obtained from coconut in which the long chains are C8 (caprylic) and C10 (capric) respectively. The chemical names of the corresponding acids are octanoic acid and decanoic acid. This combination is widely used as an emollient; it softens the skin and yet does not feel greasy.

**Caprolactam** is azepan-2-one and is a cyclic compound from which nylon is made. It can be added to form cross-links between polymers.

**Caramel** is made by heating sugar until it melts and turns brown. Another type of caramel is made by heating sugar with sulfur dioxide. It is more stable and has the food code E150b. Ammonia caramel, also known as confectioner's caramel, can be made by heating various sugars with ammonium hydroxide to give a dark brown colour; its food code is E150c.

**Carbamic acid butyl-3-iodo-2-propynyl ester** – see 3-iodo-2-propylbutyl carbamates.

**Carbomer** is the polymer also known as poly(acrylic acid). It is added to make a product form a smooth gel.

**Carboxymethyl ethers of starch (and cellulose)** are impervious to breakdown by chemical or microbial attack. They are used in detergents, wallpaper pastes, toothpastes, pharmaceuticals, and some foodstuffs. Carboxymethyl cellulose has food code

E466, and E468 when it is cross-linked and known as croscarmellose. There is also a form with food code E469 which
has been modified by enzymes to make it water-soluble and is
suitable as a thickener in foods.

**Carotene** consists of a C18 hydrocarbon chain with alternate
double bonds and with a six-membered hydrocarbon ring at
both ends. There are four methyl groups along the chain and
three attached to each ring. Carotene is produced by plants like
carrots (hence its name), sweet potatoes, mangoes, and apri-
cots. It is used as a food colourant and has the food code E160.

**Carrageenan** has the food code E407 and is a polymer of the
carbohydrate galactose. It is extracted from seaweed and used
to thicken products such as cosmetics, shampoos, sexual lubricants, toothpaste, and shoe polish.

**Caustic soda** is sodium hydroxide, NaOH.

**Cellulose** is a long chain polymer composed of glucose units in a
structure that is resistant to chemical decomposition. It is
extracted from wood pulp. **Cellulose gum** is a modified form of
cellulose which is stable in water and which acts as a thickening agent. It is approved for use in foods and has the food
code E466. See also **methylcellulose**.

**Cera microcristalina** is a hydrocarbon wax having 35 or more
carbon atoms in its molecular structure.

**Ceteareth-20** is a long chain polymer which incorporates oxygen
atoms along its chain, so it is technically an ether. It acts as a
non-ionic surfactant.

**Cetearyl alcohol** has long hydrocarbon chains, C15 to C17, and is a thickener and an emollient.

**Cetrimide** is cetyltrimethylammonium bromide and is a powerful antiseptic.

**Cetyl alcohol** is 1-hexadecanol and it consists of a C16 hydrocarbon chain with an alcohol (OH) group at one end. It is used both as an emollient and to make a liquid look opaque. It can also act as a thickener of liquids and creams.

**Cetyl dimethicone copolyol** – see silicones.
**Chelating agent (chelant)** – see Technical Words.
**Chlorhexidine** consists of a pair of biguanide units to each of which is attached a chlorophenyl group. It is generally used as the cation and this makes it attractive to microbial cells walls that are negatively charged and thereby disrupts them. It is used in the form of **chlorhexidine gluconate**, which is especially deadly to bacteria. It is also active against viruses.

**4-Chloro-m-cresol** is 4-chloro-3-methylphenol, also known as *para*-chloro-methyl-cresol or PCMC. It has antifungal properties.

**Copper complexes of chlorophylls** (food code E141) can be mixed with carotene (food code E160a) to provide a pale green/yellow food colourant.

**Citric acid** is 2-hydroxypropane-1,2,3-tricarboxylic acid. It has a fruity acid tang and is naturally present in lemons and limes. It is manufactured industrially by the action of the *Aspergillus niger* fungus on sugar, and around 1.5 million tonnes are made every year. Its food code is E330.

**Citronellol** is the main fragrance of geraniums and roses, and is 3,7-dimethyloct-6-en-1-ol. It can exist as a mixture of left and right hand molecules, and it is the less common right hand molecule that provides the fragrance.

**Clopyralid** is the trade name of the herbicide 3,6-dichloro-2-pyridinecarboxylic acid.

**Cocamide MEA** is $CH_3(CH_2)_nCONHCH_2CH_2OH$, where $n$ indicates that the fatty chain which comes from coconut oil is of indeterminate length. It is a non-ionic surfactant and is an excellent foaming agent. The MEA stands for mono-ethanolamime – see ethanolamine.

**Cocamidopropyl betaine** is a non-ionic surfactant made from the fatty acids of coconut oil to which is attached dimethylamino-propylamine.

**Cocamidopropyl PG-dimonium chloride phosphate** is a cationic surfactant. It is used in some cosmetics because it leaves a long-lasting protective layer on skin. It is a triester of phosphoric acid, with *N*-(2,3-dihydroxypropyl)-*N*,*N*-dimethyl-3-(1-oxococo-alkyl)amino-1-propanaminium chloride groups.

**Cochineal** has the food code E120 and is obtained from the female scale insect (*Dactylopium coccus*), which lives on cacti. The red component is carminic acid, which is extracted with a solution of ammonia or sodium bicarbonate. Cochineal is again being used as a food colourant as part of the move towards using natural rather than synthetic colourants.

**Coconut oil** is extracted from the flesh of the coconut and is 90% saturated fat so does not easily go rancid. Its main fatty acids are lauric acid (dodecanoic acid, C12) and myristic acid (tetradecanoic acid, C14).

**Cocoyl caprylocaprate** consists of saturated fatty acids with 8 or 10 carbons in their chain, linked to fatty alcohols with 12 to 18 carbons in their chains.

**Colour index (CI).** A number is given to every dye that has been approved for use, and those used in foodstuffs are also given an E-food code as well.

**Cranberries** owe their flavour to the molecules ethyl-2-methyl-butyrate, ethyl-3-methylbutryate, and *trans*-2-hexenal.

**Crotamiton** is *N*-ethyl-*N*-(2-methylphenyl)but-2-enamide. It can be bought over-the-counter as an antipruritic (anti-itching) treatment.

**Cyclamate** is the anion of cyclamic acid, which is cyclohexyl-sulfamic acid.

**Cyclodextrin,** also known as α-cyclodextrin, is a water-soluble carbohydrate made by enzymes from starch, and it exists as a ring of six glucose units joined together. It has a band-like structure and is able to wrap around smaller molecules and ions.

**Cyclopentsiloxane** – see silicones.

**Decanoic acid, sodium salt** is also called sodium caprate, and has the formula $CH_3(CH_2)_8CO_2Na$. It is an anionic surfactant and attacks grease aggressively.

**Decyl oleate** and **glyceryl oleate** are ester derivatives of oleic acid, and this acid comes mainly from olive oil. The former is the ester formed with 1-decanol, and the latter with polyalcohol glycerol.

**Dextrose** is another name for glucose, more correctly called D-glucose. The D is short for dextrorotatory (*dextra* = right

handed) to distinguish it from the left handed form. D-Glucose
is the one produced naturally by plants and which is easily
digested.

**4,5-Dicholor-2-octyl-2H-isothizol-3-one** is a powerful biocide.

**Diclofenac diethylammonium** is a non-steroid anti-inflammatory
drug (NSAID) and offers relief by blocking certain enzymes.

**Dicocoyl pentaerythrityl distearyl citrate** is a complex material
made from the natural fatty acids derived from coconut
oil, which are stearic acid and citric acid, combined with the
industrial chemical pentaerythritol. It is used to soften
the skin.

**Diethanolamine** – see ethanolamine.

**Diethylamine** is $(CH_3CH_2)_2NH$ and is used in the manufacture of
rubber, dyes, pesticides, and pharmaceuticals.

**Diethylene glycol** has the formula $CH_3OCH_2CH_2OCH_2CH_2OH$
while **diethylene glycol monoethyl ether** has an ethyl group in
place of the $CH_3$ group. See also glycol ethers.

**Dihydroxyacetone** is $HOCH_2COCH_2OH$ and is a white crystalline
powder which is prepared industrially by the fermentation of
glycerol (glycerine) by *Acetobacter suboxydans.* It is used to
darken the skin.

**Dilauryl thiodipropionate** is an antioxidant.

**Dimethicone** – see silicone. (Dimethicone is short for dimethylsilicone.)

**Dimethicone copolyol** – see silicone.

**Dimethyl dicarbonate** is $CH_3OC(O)–O–C(O)OCH_3$. This molecule acts as a preservative, especially for drinks, by disabling enzymes in microbes like yeasts which cause spoilage. It is now being added to wines in place of sulfur dioxide to which some people are sensitive.

**Dimethyl ether** is $CH_3OCH_3$ and it is a gas with a boiling point of $-24\ °C$. It is used as an aerosol propellant.

**Dimethyl naphthalene** is a hydrocarbon solvent which has two co-joined benzene rings to which are attached two methyl ($CH_3$) groups, variously arranged. This liquid boils at around $260\ °C$ and acts both as a solvent and as a lubricant.

**Dipotassium phosphate** is $K_2HPO_4$. It has the food code E340, and is used as a raising agent and as a buffering agent.

**Disodium inosinate** is the sodium salt of inosinic acid and is used to enhance the umami flavour in the same way as monosodium glutamate (MSG). It has the food code E631.

**Disteardimonium hectorite** is a combination of distearyldimonium chloride (*N,N*-dimethyl-*N*-octadecyl chloride) and the mineral

hectorite, which is a magnesium lithium silicate mineral. It is
used as a thickening agent.

**DMDM hydantoin** (DMDM stands for dimethylol dimethyl) is a
powerful antimicrobial agent which works by releasing
formaldehyde.

**E numbers** – see Technical Words.
**Emollient** – see Technical Words.
**Emulsifier** – see Technical Words.
**Erythrosine** is an intense red dye, also known as Red No.3. It is
2,4,5,7-tetraiodofluorescein and can be used to colour food. It
has the food code E127.

**Ester** is the generic name for the product of the reaction of an
acid ($R^1CO_2H$) and an alcohol ($R^2OH$) and has the formula
$R^1CO_2R^2$. Esters are noted for their fruity aromas and many
fruits contain them.
**Ethanolamine** is also known as 2-aminoethanol and it comes in
three forms: monoethanolamine (MEA, pictured), diethanolamine (DEA), and triethanolamine (TEA) depending on how
many ethanol units ($-CH_2CH_2OH$) are attached to the nitrogen. MEA is used as a buffer to keep the pH slightly alkaline.
They also have mild surfactant properties.

**Ethoxydiglycol** is $CH_3CH_2OCH_2CH_2OCH_2CH_2OH$. It is a gentle solvent and suitable for products that come in contact with the skin.

**Ethoxylated alcohols** – see alcohol ethoxylate.

**Ethoxylated tallow amine** is a surfactant – see tallow.

**Ethyl acetate** is $CH_3C(O)OCH_2CH_3$. It is a volatile liquid used as a solvent and it has a boiling point of 77 °C.

**Ethylcellulose** is cellulose in which some of the OH groups have been converted to $OCH_2CH_3$. It is used as a slow-release additive in tablets.

**Ethylene glycol** is ethane-1,2-diol with the formula $HOCH_2CH_2OH$. Its most common use is as an antifreeze.

**Ethylene glycol monostearate** acts as an emollient and thickening agent for creams.

**Ethylene/methacrylate copolymer** is used to form a protective film on a surface.

**Ethylhexyl salicylate** is also known as octyl salicylate and is 2-ethylhexyl 2-hydroxybenzoate. It acts as a UV filter.

**Ethyl methyl ketoxime** is $(C_2H_5)(CH_3)C=NOH$. It is used to prevent a skin forming on the surface of a tin of paint.

**Etidronic acid** is 1-hydroxyethane-1,1-diphosphonic acid. It is a chelating agent, able to bind to calcium and thus soften water. It is used in detergents and cosmetics as it also suppresses the formation of free radicals, such as those produced by UV.

**Eucalyptol** is 1,3,3-trimethyl-2-oxabicyclo[2,2,2]octane. It is obtained by distilling eucalyptus oil. It has medicinal benefits such as reducing pain and easing breathing.

**Excipient** – see Technical Words.

**Fatty acids** can be either saturated, monounsaturated with one double bond along the chain, or polyunsaturated, with two or more double bonds. The ordinary names of the more common saturated ones are caprylic acid (C8), capric acid (C10), lauric acid (C12), myristic acid (C14), palmitic acid (C16) and stearic acid (C18). They are more correctly called octanoic, decanoic, dodecanoic, tetradecanoic, hexadecanoic, and octadecanoic acids, respectively. Of those produced in Nature, the main monounsaturated one is oleic (C18) and the common polyunsaturated one is linoleic (also C18).

**Fatty alcohol alkoxylate** consists of long hydrocarbon chains (fatty alcohol), as in alcohol ethoxylates.

**Ferrous sulphate** is $FeSO_4$, and is more correctly called iron(ɪɪ) sulfate.

**Fipronil** is 5-amino-1-[2,6-dicholor-4-(trifluoromethyl)phenyl]-4-(trifluoromethylsulfinyl)-pyrazole-3-carbonitrile. It is an insecticide.

**Fixative** – see Technical Words.

**Fluorphlogopite** is an aluminium silicate fluoride mineral.

**Fluroxypyr** is the herbicide [(4-amino-3,5-dichloro-6-fluoro-2-pyridinyl)oxy]acetic acid.

**Folic acid** is also known just as folate or vitamin $B_9$.

**Formaldehyde** is methanal and has the formula $H_2CO$.

**Fragrance molecules** have to be identified if they are part of a perfume raw material that has been added to a product. The ones so identified, and the smells associated with them, are as follows:

**Amyl cinnamal** smells like jasmine.

**Butylphenyl methylpropional** is just a flowery smell (see alphabetical entry for structure).

**Citral** smells of lemon.

**Citronellol** smells of geraniums and roses – see formula under citronellol.

**Coumarin** has the scent of new mown hay.

**Eugenol** has a spicy smell like cloves.

**Geraniol** smells of roses – see geraniol.
**Hexyl cinnamal** smells of camomile.

**Hydroxyisohexyl 3-cyclohexene carboxaldehyde** smells like lilies.

**Hydroxycitronellal** smells of lime.

**α-Ionone** has a sweet violet odour, and is often said to provide a
clean outdoor smell.

**Isoeugenol** smells of vanilla.

**Limonene** smells of oranges and is extracted from orange peel.

**Linalool** has a floral, spicy aroma (see alphabetical entry for
structure).
**Methyl-2-octynoate** smells of violets.

**1-(2,6,6-Trimethyl-3-cyclohexen-1-yl)-2-buten-1-one** has a floral odour with woody undertones.

**3-(4-*tert*-Butylphenyl) propionaldehyde** smells of lily of the valley. It is also known as bourgeonal.

**Fructose** is also known as fruit sugar. It is the carbohydrate that provides the sweetness of honey and treacle, and is almost twice as sweet as sugar.

**Fumaric acid** is $HO_2CCH=CHCO_2H$. It has a fruity taste and has the food code E297. It is used to provide an acid taste.

**Gel** – see Technical Words.

**Gelatin** (gelatine) is the protein of collagen and is extracted from animal products like skin and hoofs. It is used as a thickener for water-based mixtures and for jellies. Its food code is E441.

**Geraniol** smells of roses and occurs in other plant extracts as well as rose oil. Products with high levels of geraniol are reputed to be effective mosquito repellents.

**Gluconate** is the negative ion of gluconic acid.

**Glucosamine sulfate.** Glucosamine (pictured) is glucose in which an OH group has been replaced by an amine ($NH_2$) group.

Commercially it is produced from shrimp shells and is believed to benefit joints by repairing cartilage.

**Glucose** is the simplest and most abundant carbohydrate produced by plants, which convert it to the natural polymers starch and cellulose.

**Glucose syrup** is produced from maize starch (corn starch) by the action of two enzymes. *Amylase* enzyme breaks the long chains of starch into smaller strands and then *glucoamylase* chops them into glucose molecules.

**Glutamic acid** is an amino acid abundant in fish, milk, and eggs. It has the food code E620.

**Glutaral** is also known as glutaraldehyde and is 1,5-pentanedial. It has disinfectant properties.

**Glycerin**, also known as glycerine and glycerol, is propane-1,2,3-triol. It is used as a thickening agent and humectant and has the food code E422.

**Glycol ethers** consist of short chains of the formula $R(OCH_2CH_2)_nOH$ where R can be an alkyl group, such as methyl, ethyl, or propyl, or a phenyl group. They are useful solvents and can dissolve carbohydrates.

**Glyphosate** is *N*-(phosphonomethyl)glycine. Its isopropyl-ammonium derivative is the weedkiller Roundup. (Isopropyl-ammmonium is the positive ion, $(CH_3)_2CHNH_3^+$.)

**Guaifenesin** is (*RS*)-3-(2-methoxyphenoxy)propane-1,2-diol. It is an expectorant.

**Guar gum** consists of chains of thousands of mannose molecules with galactose side groups. It is water-soluble and is excellent as a thickener for ice creams, soups, mayonnaise, toiletries, and cosmetics. Its food code is E412.

**2,2′,2″-(Hexahydro-1,3,5-triazine-1,3,5-triyl)triethanol**–see 2,2′,2″-(1,3,5-triazine-1,3,5-triyl)triethanol.

**Hexyl laurate** is the C6 ester of the C12 fatty acid, lauric acid, whose chemical name is dodecanoic acid. It is used as an emollient.

**Hexylresorcinol**, also known as 4-hexylresorcinol, is 4-hexylbenzene-1,3-diol. It is used as a skin disinfectant and is an acceptable antioxidant food additive with food code E586.

**Homosalate** is 3,3,5-trimethylcyclohexyl 2-hydroxybenzoate. It is used as a sunscreen because it absorbs UV rays.

**Humectant** – see Technical Words.

**Hydrated silica**, also known as silicic acid, approximates to $H_2SiO_3$. It can be used as an abrasive, an anticaking agent, to bulk things out, and as an absorbent.

**Hydrogenated vegetable oils** are made by treating unsaturated fatty acids with hydrogen, thereby making them saturated and more stable because they are less likely to be oxidised.

**Hydroxyethyl cellulose** is cellulose in which some of the hydrogens of its OH groups have been replaced by ethyl ($CH_2CH_3$) groups. It is used to improve viscosity and lubricity.

**Hydroxypropyl methylcellulose** is a polymer based on modified cellulose with both methyl groups and propyl groups attached, with these having OH or $OCH_3$ groups incorporated. It is viscous and is used as a thickening agent. Its food code is E464.

**8-Hydroxyquinoline**, also known as quinolin-8-ol, is a powerful antibacterial.

**Hypromellose** – see hydroxypropyl methylcellulose.

**Imidazolidinyl urea** is a preservative which releases formaldehyde as an antimicrobial agent.

**Inosinic acid** and **guanylic acid** are used to make the flavour enhancer known as disodium 5′-ribonucleotide. Isosinic acid is also known as inosine monophosphate (IMP), while guanylic acid is also known as 5′-guanidylic acid (GMP).

**Inositol** is a simple carbohydrate which has a role as a messenger molecule and plays a part in the nervous system and in the breakdown of fats to release energy. It is cyclohexane-1,2,3,4,5,6-hexol.

**3-Iodo-2-propylbutyl carbamate** is a biocide and is $CH_3(CH_2)_3NHC(O)OCH_2C\equiv CI$.

**Iron oxides** are of three types and three shades: there is black iron oxide, $Fe_3O_4$ (which is also known as magnetite), yellow iron oxide, $HFeO_2$, and brown ferric oxide $Fe_2O_3$.

**4-Isopropenyl-1-methylcyclohexene** is limonene – see fragrance molecules.

**Isopropyl alcohol** is $(CH_3)_2CHOH$ and is a solvent of boiling point 83 °C.

**Isopropylammonium glyphosate** – see glyphosate

**Lactic acid**, also known as milk acid, is $CH_3CH(OH)CO_2H$. It is also present in bread, cheese, meats, beers, and wines. Its food code is E270. It is also a component of human skin and essential for keeping it healthy.

**Lactose** is also known as milk sugar and it consists of two linked carbohydrate molecules, glucose and galactose. It occurs naturally in milk and especially in human milk, of which it comprises 7% – more than in any other animal's milk.

**Lanolin** is a mixture of long chain fatty acids linked to long chain alcohols. It is excreted by the hair follicles of sheep and makes their wool waterproof. It is often used in ointments for cracked skin, chapped lips, and sore nipples.

**Lanolin alcohol** is a mixture of organic alcohols obtained by the hydrolysis of lanolin, and it is used as an emollient, emulsifier, and antistatic agent.

**Lauramidopropyl betaine** is a gentle surfactant produced sustainably from coconut oil and sugar beet. It is an amphoteric surfactant.

**Lauramine oxide**, also known as lauryl dimethylamine oxide, is $CH_3(CH_2)_{11}N(O)(CH_3)_2$. It is produced from plant oils and is a non-ionic surfactant.

**Laureth-4** is a non-ionic surfactant with subsidiary roles as an emulsifier and thickener. It is $CH_3O(CH_2CH_2O)_4OR$ where R is the C12 hydrocarbon group known as lauryl or dodecyl. There are other laureths such as **laureth-7**, **laureth-8**, **laureth-10**, and **laureth-20** and these are sometimes referred to as PEG lauryl ethers.

**Lauric acid** is dodecanoic acid and is a C12 saturated fatty acid. It can be used as a surfactant.

**Lecithin** consists of glycerol, with two fatty acid ester chains and a phosphate, linked to choline. It has applications in pharmaceuticals, plastics, inks, and paints, and acts as an emulsifier and dispersant. Its food code is E322. Lecithin is extracted from animal tissues and from soya beans.

**Leucine** is an essential amino acid, meaning it has to be part of our diet since we cannot synthesise it within the body. It is present in soya, maize, and wheat and has the food code E641. It is $(CH_3)_2CHCH_2CH(NH_2)CO_2H$.

**Levonorgestrel** is the active agent in some morning-after pills.

**Linalool** is 3,7-dimethylocta-1,6-dien-3-ol and can exist as mirror images, both of which are produced in Nature. *S*-Linalool has a coriander aroma, while *R*-linalool smells like lavender.

**Liquid paraffin** is a mixture of hydrocarbons distilled from crude oil and it is a collection of similar molecules with 15–40 carbon atoms in their chains.

**Lithium** is the lightest of all metals, atomic number 3, and its compounds are used in batteries, greases, and medicine to treat bipolar depression.

**Locust bean gum** is a powder extracted from the seeds of the carob tree and it is used as a gelling agent. It consists of long chain carbohydrates made up of galactose and mannose units. Its food code is E410.

**Locust bean hydroxypropyltrimonium chloride** has a 2-hydroxy-*N,N,N*-trimethyl group attached to a carbohydrate chain and is used in cosmetics as a hair or skin conditioner.

**Loperamide** is 4-[4-(4-chlorophenyl)-4-hydroxypiperidin-1-yl]-*N,N*-dimethyl-2,2-diphenylbutanamide. It will stop diarrhoea

by blocking receptors that cause the normal mode of action
that moves the gut's contents along.

**Lysine** is $NH_2CH_2CH_2CH_2CH(NH_2)CO_2H$ and is an essential
amino acid, meaning it has to be part of our diet since we
cannot synthesise it within the body. It is present in red meat,
eggs, beans, fish, and soya.

**Macroglycerol hydroxystearate** is also known as PEG-40 castor oil
(and has several commercial names). It is made by reacting
castor oil, that has been hydrogenated to remove its double
bonds, with ethylene oxide, which adds a polymer chain to the
oil. It is used as an emulsifier and also used to make aqueous
solutions of ingredients which would otherwise not be miscible
with water, such as vitamins A, D, E, and K.

**Macrogol cetostearyl ether** consists of polyethylene glycol to
which are attached fatty alcohols.

**Magnesium stearate** is magnesium octadecanoate. It is used
when manufacturing tablets to prevent them from sticking to
the stamping machinery. It is also used as an anti-caking
agent. Its food code is E470b.

**Magnesium trisilicate** is $Mg_2Si_3O_8$, and it is used as an abrasive,
an absorbent, a bulking ingredient, an anti-caking agent, to
control viscosity, or to make something opaque. It also can act
as a glidant – in other words it is added to powders to make
them flow better.

**Malic acid** is the sour taste of unripe fruits. It exists in both left
and right hand forms of which Nature produces only the former while that produced by the chemical industry contains
both. Malic acid is used in chemical peels in which the top
layer of skin is removed to reveal more youthful looking skin.
Its food code is E296.

**Maltitol** is a disaccharide made from starch. It is almost as sweet as sugar but unlike sugar it does not cause tooth decay. It is 4-*O*-α-D-glucopyranosyl-D-glucitol and has food code E965.

**Maltodextrin** is another name for glucose syrup and it consists mostly of short chains of linked glucose molecules. It is made from wheat starch.

**Mannitol** has the same chemical formula as sorbitol (hexane-1,2,3,4,5,6-hexol) but a slightly different molecular structure. It has similar sweetness. Its food code is E421.

**MEA** stands for methylethanolamine, chemical formula $H_3CNHCH_2CH_2OH$ – see also ethanolamine.

**MEA citrate** is a salt of citric acid and it acts as a sequestering agent, especially for the calcium ions of hard water.

**MEA palm kernelate** is a sustainable surfactant made from palm oil and its long chain component is mainly C16.

**Melanoidins** are polymers formed by the reaction of carbohydrates and amino acids (sugars and proteins) and are dark brown.

**Menthol** is 5-methyl-2-(1-methylethyl)cyclohexanol. It occurs naturally in peppermint oil and is an ingredient in throat lozenges and cooling salves. It has a cooling effect because it can triggers cold-sensitive receptors in the skin, the nose, or the throat.

**Metaldehyde** is 2,4,6,8-tetramethyl-1,3,5,7-tetraoxacyclo-octane and is used to poison slugs.

**Methacrylate, methyl acrylic/acrylic acid copolymer modified** – see acrylic acid.

**Methenamine**, also known as hexamethylenetetramine, has the formula $(CH_2)_6N_4$. It is an intermediate in the manufacture of various plastics and pharmaceutical drugs, but is also a drug in its own right, especially against bladder infections when it is prescribed as the salt of mandelic acid. Used as a food preservative, its food code is E239. It decomposes in the presence of acid to release formaldehyde, which is a powerful antibiotic agent.

**Methicones** – see silicones.

**Methyl-1*H*-benzotriazole** is used as a corrosion inhibitor. It forms a protective film on the surface of metals like copper and silver and protects them.

**Methylcellulose** is cellulose in which methyl groups have replaced one, two, or all three of the hydrogen atoms of the OH groups of the glucose molecules that comprise the polymer cellulose. It is used as an emulsifier and thickener in foods and has the food code E461. It is used in cosmetics, wallpaper paste, and even as a treatment for constipation. Methylcelluose is graded according to the viscosity it produces.

**Methylchloroisothiazolinone** is methylisothiazolinone to which a chlorine atom has been attached, making it a stronger antimicrobial agent.

**2-Methyl-4-chlorophenoxyacetic acid iso-octyl ester** is a weedkiller.

**Methyl hydroxybenzoate** and **propyl hydroxybenzoate** are powerful preservatives often simply referred to as parabens – see parabens – and they are added to foods because they counteract bacteria and fungi. Their food codes are E218 and E217, respectively, and they are the sodium salts of the methyl and propyl esters of hydroxybenzoic acid.

**Methylisothiazolinone** (MIT) is a powerful biocide and is used to keep personal care products, like hand lotions and shampoos, free of germs.

**2-Methylpentane-2,4-diol** acts as an emulsifier.

**Methyl salicylate** is better known as oil of wintergreen and is methyl-2-hydroxybenzoate. It is used as a rubbing oil to treat aches and pains.

**Microcrystalline cellulose** – see cellulose.

**Microcrystalline wax** is obtained from oil and consists of long hydrocarbon chains with side chains. It is used to provide solid support to other materials. It melts in the range 60–80 °C.

**Miglyol 812** comes from coconut oil and is a mixture of octanoic (caprylic, C8) and decanoic (capric, C10) triglyceride. It is used as an emollient to soften the skin.

**Milk protein** (and **milk protein concentrate**) is mainly casein and is a rich source of amino acids, carbohydrates, phosphate, and calcium. It is left behind when other components of milk, namely lactose, minerals, and water, have been filtered out. It is used in processed foods.

**Mineral oil** – see liquid paraffin.

**Minoxidil** was designed as a drug to reduce blood pressure but is now used to stimulate hair growth. It is 6-piperidin-1-yl-pyrimidine-2,4-diamine 3-oxide.

**Modified starch** is starch that has been treated to make it more soluble in cold water, and this is done by heating an aqueous suspension and then drying it, a process known as pre-gelatinization. Starch can also be modified by reaction with ethylene oxide or propylene oxide to make thickeners. It has food code E1401.

**Mono- and di-glycerides of fatty acids** are emulsifiers and are added to ensure that ingredients blend to a smooth paste. Their food codes are E471 and E472, respectively.

**Monoethanolamine** – see ethanolamine.

**Monosodium glutamate** (MSG) is $NaO_2CCH(NH_2)CH_2CH_2CO_2H$. It is an amino acid and is present naturally in the human body. It is part of the basic taste sensation known as umami or savoury. It has the food code E621.

**Mordant** – see Technical Words.

**MSG** is monosodium glutamate.

**Myreth-3 myristate** is tetradecoxyethyl tetradeconoate, and it is used as an emollient, skin conditioner, and mild surfactant.

**Myristyl alcohol** is 1-tetradecanol, and it is used as an emollient in personal care products.

*N,N*-**Dimethyltetradecylamine** *N*-**oxide** is used to boost the action of surfactants. In personal care products it is used as an antistatic and emulsifying agent.

**Niacin,** also known as nicotinic acid or vitamin $B_3$, is pyridine-3-carboxylic acid. It assists enzymes in releasing energy from carbohydrates. Fish, poultry, and nuts contain a lot.

**Nitrocellulose** is cellulose to which nitro groups $(NO_2)$ have been attached to the glucose molecules that make up this natural polymer. It is a highly flammable compound that was once used to make plastics (celluloid) and cine film. Dissolved in ethanol it is known as collodion and is used as a lacquer (also known as dope).

**Octocrylene** is 2-ethylhexyl 2-cyano-3,3-diphenyl-2-propenoate and is used in sunscreen to protect against UV rays.

**Octadecane-1-thiol** consists of a hydrocarbon chain of 18 carbons with a thiol group (SH) at the end of the chain. It is used to clean and protect silverware.

**Octyldodecanol** is a branched long chain fatty alcohol. It is used in personal products as an emulsifier and to makes a liquid appear opaque.

**Oleth-2** and **oleth-5** are emulsifiers and non-ionic surfactants and are used in personal care products. They are the polyethylene glycol ethers of oleyl alcohol with the 2 and 5 signifying the ratio of the ether to the alcohol.

**Oleyl alcohol** is octadec-9-en-1-ol and is a C18 chain with a double bond, making it an omega-9 monounsaturated alcohol. It is used as an emulsifier and thickener for shampoos as well as acting as an emollient.

**Oligofructose** is a carbohydrate consisting of a short chain of linked fructose units. It is not broken down by enzymes in the stomach so it is not absorbed, but provides food for beneficial bacteria in the intestines.

**Oxalic acid** is ethanedioic acid and is a solid. It is used as a cleaning agent to remove limescale and microbes.

**Panthenol** is 2,4-dihydroxy-*N*-(3-hydroxypropyl)-3,3-dimethylbutan-amide and is similar to vitamin B5. It acts as humectant and is used in skin and hair products.

**Pantothenic acid** is 3-[(2,4-dihydroxy-3,3-dimethylbutanoyl)amino]-propanoic acid, and is needed by the body to release energy from stored fats.

**Parabens** are esters of *para*-hydroxybenzoic acid and named as such. They have the chemical formula $HOC_6H_4CO_2R$, in which R can be methyl $(CH_3)$, ethyl $(C_2H_5)$, or propyl $(C_3H_7)$ – the choice of which to use depends on the solubility required for it to blend with other ingredients.

**Paraffin** in ingredient lists is not the liquid fuel commonly called by this name (which is also known as kerosene) and which consist of hydrocarbons of length C6 to C15. Cosmetic **paraffin** refers to a mixture of hydrocarbons of much longer chains, some of which are waxes when in a pure state. They are straight chain hydrocarbons of C20 up to C50.

**C9–11 Pareth-6, C11-15 pareth-3, C11–15 pareth-9, C12–14 pareth-7, C12–15 pareth-3,** and **C14–15 pareth-7** are non-ionic surfactants. They consist of chains of carbon atoms of different lengths between 9 and 15 and indicated by the **C** prefix numbers. They are attached to a chain made up of 3–9 ethylene oxide $(CH_2CH_2O)$ units, this being the number given as a suffix. Sometimes the length of the chain is not indicated and the term **pareth-n** is used. They are surfactants.

**PEG, PEG-6, PEG-8** *etc.* stand for polyethylene glycol, whose general formula is $HO(CH_2CH_2O)_nH$, where *n* can be incorporated into the name – as in **PEG-6**, although the 6 is an

average and other PEGs are likely to be present, such as PEG-5
and PEG-7. These compounds are used as humectants.

**PEG-45/dodecylglycol copolymer** is made by the reaction of the
naturally occurring fatty acid lauric acid and ethylene oxide. It
is used to stabilise emulsions. **PEG-23M** is a foam booster.

**PEI** stands for polyethylene imine. It consists of chains of carbon
atoms interspersed with nitrogen atoms.

**Pentaerythrityl tetraisostearate** is the tetra ester of pentaerythriyl
with iso-stearic acid and is used as an emollient.

**Pentasodium triphosphate** is $Na_5P_3O_{10}$ and it is a water softener,
which it does by chelating calcium ions and so preventing
them interfering with surfactants.

**Perfume raw material** – see fragrances.

**Permethrin** is an insecticide of the pyrethroid kind.

**Phenethyl alcohol** is 2-phenylethanol and a powerful antimicrobial
agent. It is used as an antimicrobial, antiseptic, and disinfectant.

**Phenoxyethanol** is 2-phenoxyethanol and is $C_6H_5OCH_2CH_2OH$. It is a preservative in personal care products.

**Phenylalanine** is $C_6H_5CH_2CH(NH_2)COOH$ and is an essential amino acid for humans – in other words, we must get it from the food that we eat. However, there are some people who suffer from the genetic disorder phenylketonuria which converts some of it to phenylketone, which is toxic.

**Pholcodine** is morpholinylethylmorphine and manufactured from morphine by adding a cyclic morpholine group $(-NC_4H_8O)$ to the molecule. It is an over-the-counter medicine for persistent coughs.

**Phosphoric acid** is $H_3PO_4$ and is used to clean metal surfaces, and especially to remove rust. It leaves behind a protective phosphate layer. It is also used in fizzy drinks like colas. It has the food code E338.

**Phosphorus** is chemical element number 15 and it can be either the red form $(P_n)$ which is stable, or white phosphorus $(P_4)$, which is dangerously flammable.

**Plant sterol esters** are also known as phytosterols. They are chemically similar to cholesterol and can block the cholesterol-absorbing sites in the intestines. They are extracted from pine tree oil and made soluble in oils and fats by attaching a long chain fatty acid.

**Poloxamer 407** is a copolymer of propylene oxide and ethylene oxide. It is used to solubilise oily ingredients. Poloxamers are non-ionic surfactants used as defoaming agents, wetting agents, and emulsifiers.

**Polycarboxylate** is a polymer of acrylic acid, sometimes copolymerized with maleic acid. It is often used in the form of the water-soluble sodium salt as a thickening agent, working best under slightly alkaline conditions. See also carbomer.

**Polyglyceryl-4 isostearate** consists of glycerol (1,2,3-propanetriol) with ester links to branched C18 isostearic acids (iso-octadecanoic acid). It acts as an emulsifier.

**Polymeric biguanide hydrochloride** is also known as poly-hexamethylene biguanide. It is an ionic polymer with anti-microbial action and used as a preservative.

**Polyoxyl 40 stearate** is a non-ionic surfactant and the 40 refers to the number of repeating units in its polymer chain of $-OCH_2CH_2-$ molecular segments. This chain is attached to stearic acid, which has 18 carbons.

**Polyquaternium** is an internationally agreed term for describing cationic polymers used in personal care products and it signifies that the positive centres are quaternary ammonium units. When a number follows this name it merely identifies when the polymer was registered. **Polyquaternium-7** and **polyquaternium-39** are made from acrylic acid, acrylamide, and diallyldimethylammonium chloride, and both are used as moisturisers.

**Polysorbate 60** is also known as polyoxyethylene sorbate and is used as an emulsifier.

**Polyvinylpyrrolidone** – see PVP.

**Potassium carbonate** is $K_2CO_3$ and is used to generate bubbles when in contact with an acid.

**Potassium chlorate** is $KClO_3$ and is a powerful oxidising agent.

**Potassium chloride** is KCl and is the potassium equivalent of common salt (sodium chloride). Like salt, it is essential for the working of the nervous system. Its food code is E508.

**Potassium nitrate** is $KNO_3$ and is a component of gunpowder. It is also known as nitre.

**Potassium sorbate** is potassium hexa-2,4-dienoate and is a preservative used in foods and drinks; food code is E202. It is an effective fungicide which interferes with the reproductive capabilities of moulds and yeasts. It occurs naturally in rowan berries.

**Potassium sulfate** is $K_2SO_4$ and it is used as a fertilizer.

**Povidone** – see PVP.

**PPG** stands for polypropylene glycol – see PEG.

**Preservative** – see Technical Words.

**1,3-Propanediol** is $HOCH_2CH_2CH_2OH$ and is a solvent derived from maize (corn). It is used in conjunction with preservatives such as phenoxyethanol.

**Propan-2-ol**, also known as isopropyl alcohol, is $(CH_3)_2CHOH$. It is used as a solvent.

**Propellants** – see Technical Words.

**1-Propenyl cysteine sulfoxide** is the molecule in onions that enzymes change to thiopropionaldehye-*S*-oxide, $CH_3CH_2CH=S=O$, which we associate with onion smell and flavour.

**Propyl acetate** is propyl ethanoate, $CH_3CO_2CH_2CH_2CH_3$, and is a solvent.

**Propylene carbonate** (PC) is 4-methyl-1,3-dioxolan-2-one and is a polar solvent. It is used as an electrolyte in lithium batteries.

**Propylene glycol** is 1,2-propanediol, $CH_3CH(OH)CH_2OH$. It is a solvent with roughly the same dissolving properties as ethanol. It is used as an antifreeze and can be added to foods as an emulsifier, where its food code number is E1520. It is also used as a viscosity modifier, humectant, and skin conditioner.

**Propylene glycol dicaprylate/dicaprate** is a mixture of esters of caprylic acid (octanoic acid, C8) and capric acid (decanoic acid, C10) linked to propylene glycol. It is used as an emollient.

**Propyl gallate** is propyl-3,4,5-trihydroxybenzoate. It is a powerful antioxidant and used to protect products from attack by atmospheric oxygen. It has the food code E310.

**PVP, povidone, polyvinylpyrrolidone** are alternative names for a water-soluble polymer which acts as an adhesive and as a thickening agent for personal care gels. It is added to viscous liquids to prevent their contents from crystallising out. Its food code is E1201.

**Pyroxylin** is another name for nitrocellulose.

**Quinine** is (6-methoxyquinolin-4-yl-8-vinylquinuclinid-2-yl)methanol. It is a natural drug that reduces fevers and relaxes muscles. It is present in the bark of the cinchona tree from which it is extracted.

**Ranitidine hydrochloride** reduces the production of acid within the stomach and is used to treat heartburn caused by gastric reflux into the oesophagus.

**Rosin** is the pitch-like material obtained from pine trees. Its main component is abetic acid and this can be converted to methyl rosinate and used to revive wood.

**Riboflavin** is vitamin $B_2$. It is produced industrially by fermentation using yeast or bacteria. It has the food code E101.

**Saccharin** – see sodium saccharin.

**Salicylic acid** is 2-hydroxybenzoic acid. Its salts are known as salicylates and they are naturally present in several foods, such as currants and raisins, raspberries, almonds, tea, and honey – foods to be avoided if you are salicylate intolerant. Aspirin is acetyl salicylic acid.

**Sclerotium gum** is a carbohydrate derived from fungi and it readily forms a gel (jelly) in water.

**SD alcohol 40** is short for specially-denatured alcohol, which is alcohol that has been made undrinkable by the addition of Bitrex® (denatonium benzoate), which is the most bitter substance known.

**Sildenafil citrate** was discovered in the UK by pharmaceutical chemists searching for a treatment for angina. The drug relieved pain by relaxing the arteries but men testing it reported unexpected and prolonged erections.

**Silica** is silicon dioxide, $SiO_2$. In a finely divided form it is used as an anti-caking agent. It has the food code E551.

**Silicones** are chemically known as siloxanes and are polymers with the repeating unit $-O-Si(CH_3)_2-$. They range from oils to semi-solids and they are used in various ways, ranging from anti-foaming agents to skin care. They have various names such as **cyclopentasiloxane** (also known as **cyclomethicone**)

and **dimethicone**

Cyclopentasiloxane has the fluidity of water, making it easy to spread, and is used in skin products. Dimethicone, also known as polydimethylsiloxane, is used to give a permanent protective layer. **Dimethicone copolyol** is a silicone polymer with other polymer chains (R) attached to the carbons of its methyl groups, either at the ends of the chain or along its backbone.

The attached polymers are either polyethylene or poly-propylene glycols and they confer the benefits of solubilising the silicone oil in water, making it ideal for shampoos and skin lotions as it spreads smoothly and evenly over hair and skin. **Cetyl dimethicone copolyol** is similar and is thought to adhere better to the skin by virtue of the attached polymers.

**Skatole** is 3-methyl-1*H*-indole and is present in human faeces, and accounts for its characteristic smell.

**Sodium C10–C14 alkyl benzenesulfonate** is a negative ion sur-factant consisting of a benzene ring to which are attached a negatively charged sulfonate group $(SO_3^-)$ and a long chain hydrocarbon group, with 10–14 carbons.

**Sodium alkyl ether sulfate** is $R(OCH_2CH_2)_nOSO_3Na$, where the alkyl group R is generally a hydrocarbon chain of around 12–15 carbons. These compounds are anionic surfactants and par-ticularly good at attacking grease.

**Sodium benzoate** is a preservative that is particularly effective under acid conditions. It occurs naturally in cranberries, and has the food code E211.

**Sodium borate** occurs as the mineral borax and is $Na_2[B_4O_5(OH)_4] \cdot 8H_2O$. It is used in conjunction with enzymes, which it stabilises. It has the food code E285.

**Sodium carbonate** is $Na_2CO_3$.

**Sodium citrate**, also known as trisodium citrate, is trisodium 2-hydroxypropane-1,2,3-tricarboxylate. It is used as an acidity regulator. It is manufactured from molasses by fermenting with the mould *Aspergillus niger*. It has the food code E331(iii).

**Sodium croscarmellose**, also known as **sodium carboxymethylcellulose**, is cellulose that has been cross-linked by means of reaction with sodium monochloroacetate. It is used as an excipient in tablets and as an emulsifier in foods. It has the food code E468.

**Sodium dehydroacetate**, also known as methylacetopyronone, is a preservative with the food code E266.

**Sodium diethylenetriaminepentamethylenephosphonate** is a derivative of $NH_2CH_2CH_2NHCH_2CH_2NH_2$ with five methylphosphate groups attached to the nitrogen atoms. It is a very gentle cleaning agent and is used for delicate fabrics and for cleaning mirrors and windows.

**Sodium dodecylbenzene sulfonate** – see sodium C10–C14 alkylbenzene sulfonate. (Dodecyl means a C12 carbon chain.)

**Sodium fluoride** is NaF. It strengthens tooth enamel by converting the apatite (calcium phosphate) of which it is composed to a stronger form called fluoroapatite.

**Sodium gluconate** is sodium 2,3,4,5,6-pentahydroxyhexanoate and is used as a chelating agent. Its food code is E576.

**Sodium hexacyanoferrate(II)** is also known as sodium ferrocyanide and may seem a rather strange food additive in that it contains six cyanides. However, these are strongly bonded to iron and unable to form hydrogen cyanide, which is a deadly poison. This compound is used as an additive for salt to keep it free-flowing and has the food code E535.

**Sodium hexametaphosphate** is $(NaPO_3)_6$. It acts to chelate metal ions and thereby acts as a stabiliser/preservative. It consists of six phosphate groups in a ring, and it was once used as a water softener. It has the food code E452(i).

**Sodium hydroxide** is NaOH and is used to make a solution alkaline.

**Sodium hydroxymethylglycinate** acts a preservative and is made from glycine, the simplest of all amino acids.

**Sodium laurate** is a salt of lauric acid and is sodium dodecanoic acid, $CH_3(CH_2)_{10}CO_2Na$. It is an anionic surfactant.

**Sodium laureth sulfate** is a sustainable anionic surfactant made from coconut oil. Laureth is the name given to a saturated hydrocarbon C12 chain extended by two or three ethoxy groups. It produces a rich lather and is gentle on the skin.

**Sodium lauroamphoacetate** is a foaming surfactant which has additional features as a conditioner, such as being an antistatic.

**Sodium lauryl sulfate** is sodium dodecyl sulfate, $CH_3(CH_2)_{11}O$ $SO_3Na$. It is a mild surfactant and is added as a cleaning agent to toothpastes and shower gels. It is made from plant oils that consist of mainly saturated fats, such as coconut oil and palm oil.

**Sodium metabisulfite** is $Na_2S_2O_5$. It can be used as a preservative and has the food code E224. It is available in the form of Campden tablets, which are added to wine, cider and beer to kill bacteria and inhibit wild yeasts.

**Sodium methyl cocoyl taurate** is a salt of taurine, which is 2-aminoethanesulfonic acid and is $HOSO_2CH_2CH_2NH_2$. It occurs naturally in the human body.

**Sodium monofluorophosphate** is $Na_2FPO_3$ and is added to toothpaste to strengthen tooth enamel.

**Sodium PCA** is short for sodium pyrrolidone carbonic acid. It is a humectant naturally present in the skin.

**Sodium polyacrylate** – see acrylates.

**Sodium saccharin** is 1,2-benzisothiazol-3-one-1,1-dioxide. It was the first artificial sweetener, and is still widely used.

**Sodium selenite** is $Na_2SeO_3$. It is used in nutritional supplements.

**Sodium silicate** is also known as water glass and is $Na_2SiO_3$. It can act both as an adhesive and to protect metal surfaces.

**Sodium xylenesulfonate** consists of a benzene ring to which is attached two methyl groups and a sulfate group. It can be used as an anti-caking additive for powders, to boost surfactant performance, or to act as an emulsifier.

**Sorbitol** is hexane-1,2,3,4,5,6-hexol. It is a natural sweetener that does not cause tooth decay and it is present in ripe apples. It can also be used as a humectant and skin conditioner. That in commercial use is manufactured from glucose. It has the food code E420.

**Sorbitan caprylate** consists of a long chain monounsaturated fatty acid, caprylic acid (octanoic acid, C8) to which is attached sorbitol. It acts as an emulsifier.

**Sorbitan oleate** consists of a long chain monounsaturated fatty acid, oleic acid (octadecanoic acid, C18) to which is attached sorbitol. It acts as an emulsifier.

**Soytrimonium chloride** is a mixture of saturated and unsaturated alkyl trimethylammoniun chlorides. It is used as a surfactant in hair colourants.

**Spearmint flavour** is $R(-)$carvone and is a natural oil produced by the plant *Mentha spicata*. This is the right handed form of the molecule. The left handed form is known as $S(+)$carvone and has the flavour of caraway seeds.

**Spinosad** is an insecticide that was found in the bacteria *Saccharopolyspora spinosa* in soil inside an abandoned rum distillery in the Virgin Islands. It is a complex molecule that kills an insect by over-stimulating its nervous system.

**Starch ether** is starch which has been treated with propylene oxide to convert it to a more stable material. It is used as a thickener for glues and plaster. Starch ether is also known as hydroxypropylated starch and has been approved for use as a food thickener, food code E1440.

**Stearalkonium** is a positively charged molecule based on nitrogen bonded to four other groups, two methyls ($CH_3$), a benzyl ($C_6H_5CH_2$), and a stearyl (octadecyl) group ($C_{18}H_{37}$). As stearalkonium chloride it is used as an antistatic agent.

**Stearalkonium hectorite** is based on the mineral hectorite, which is a magnesium lithium silicate. It is used to create a pearlescent sheen.

**Stearamidopropyl dimethylamine** is attracted to the strands of hair by virtue of its two nitrogen atoms, and then it acts as an antistatic agent. It also has biocidal properties.

**Steareth-21** is a combination of a polyethylene glycol chain with 21 repeating units, on average, with a stearyl group (C18) attached to the end of the chain. It acts as an emulsifying agent to blend all the other ingredients into a homogeneous mixture. **Steareth-20** has 20 repeating units.

**Stearic acid** is octadecanoic acid (C18) and comes mainly from animal fats. It is a saturated acid and used as an emulsifier and as a soap, including lubricating greases.

**Stearyl alcohol** is 1-octadecanol and it is $C_{18}H_{37}OH$ with the OH group at the end of a saturated hydrocarbon chain. It can be used as an emulsifier and an emollient.

**Stevia** is a natural intense sweetener extracted from the Stevia plant. Its sweetness is derived from a glycoside called stevioside, which is based on the molecule steviol to which are bonded glucose groups.

**Strontium carbonate** is $SrCO_3$. It is used to create red flares and in fireworks.

**Strontium chloride hexahydrate** is $SrCl_2 \cdot 6H_2O$.

**Strontium nitrate** is $Sr(NO_3)_2$. It is used to create red flares and in fireworks.

**Succinic acid** is butanedioic acid, $HO_2CCH_2CH_2CO_2H$. It is used as a food additive for dough because it makes it more pliable. Its food code is E363.

**Sucralose** is 1',4,6'-trichlorogalactosucrose trichlorosucrose. It is an artificial sweeter made from sugar to which three chlorine atoms have been attached. Its food code is E995.

**Sulfamic acid** is $HSO_3NH_2$ and is a strong acid. It is a crystalline solid so is easier to transport and store than liquid acids. It is ideal for tablets that need to contain an acid.

**Superabsorbent polymer**, also known as SAP, has to be able to attract and cling on to large numbers of water molecules and is used in nappies, incontinence pads, and feminine hygiene products. The polymer most used is **polyacrylate**.

**Surfactant** – see Technical Words

**Synthetic beeswax** is made from fatty acid esters, fatty acid alcohols, and hydrocarbon waxes, which are blended to give the same consistency as natural beeswax.

**Synthetic wax** consists of long chain hydrocarbons which are extracted when refining oil. The shorter chains are soft and are those used in Vaseline®, the longer ones are harder and used as candle wax. Synthetic wax can be used to increase viscosity, to stabilize an emulsion, to coat fibres and hair and thereby prevent static attraction, and to prevent moisture loss from the skin.

**TAED** is **t**etra**a**cetyl**e**thylene**d**iamine and is used in laundry formulations where it reacts with sodium carbonate peroxide to form sodium peracetate, which is the active bleaching agent.

**Talc** is magnesium silicate, formula $Mg_3Si_4O_{10}(OH)_2$, and it is a natural white mineral renowned for its softness. It also has the food code E553b and can be used as an excipient.

**Tallow** is animal fat consisting mainly of saturated fats although one of its fatty acids is oleic acid (C18), which is a mono-unsaturated fatty acid. The other acids are palmitic acid (hexadecanoic acid, C16) and stearic acid (octadecanoic acid, C18), which are saturated.

**Taurine** is 2-aminoethanesulfonic acid with the chemical formula $HOSO_2CH_2CH_2NH_2$. Taurine is needed by the body for several functions such as digestion, in blood vessels, the eyes, and white blood cells.

**TEA-hydrogenated cocoate** is triethanolamine (TEA) combined with coconut oil that has been treated with hydrogen to remove any residual double bonds. It acts as a surfactant.

**Tetrasodium EDTA** is the sodium salt of **e**thylene**d**iamine**t**etra**a**cetic acid, and it is a powerful chelating agent, targeting the calcium ions that are responsible for hard water.

**Tetrasodium etidronate** is a phosphonate and acts as a chelating agent for deactivating metal ions which might interfere with the intended use of a product.

**Thiacloprid** is 3-[(6-chloro-3-pyridinyl)methyl-2-thiazolidinylidene]-cyanamide. It is an insecticide.

**Thiopropionaldehye-S-oxide** – see 1-propenyl cysteine sufoxide.

**Thymol** is 5-methyl-2-isopropyl-1-phenol and is a white crystalline material with antiseptic properties.

**Titanium dioxide** is $TiO_2$. It is a brilliant white pigment used in paints and in confectioner's icing. Its food code is E171.

**2,2′,2″-(1,3,5-Triazine-1,3,5-triyl)triethanol** acts as a biocide and disinfectant.

**Tolylfluanid** is 1,1-dicholor-N-[(dimethylamino)-sulfonyl]-1-fluoro-N-(4-methylphenyl)methanesulfenamide. It is a powerful anti-fungus agent used to protect wood and crops like apples and grapes.

**Triacetin** is 1,2,3-triacetoxypropane. It is used as a humectant and solvent for flavourings. Its food code is E1518.

**Trideceth-9** is also known as isotridecyl alcohol polyoxyethylene ether. It is an emulsifier and wetting agent.

**Triethanolamine** (TEA) is $N(CH_2CH_2OH)_3$. It is a viscous liquid that can act as an emulsifier or, when reacted with a fatty acid, it can act as a surfactant.

**Triethylene glycol** (TEG) is $H(OCH_2CH_2)_3OH$. It is used as an air freshener.

**Triethylene glycol diamine** is used to form gels by cross-linking other molecular units.

**Triethylene glycol rosinate** is a tacky viscous liquid to which hairs adhere strongly – strong enough for them to be pulled from their roots. It is manufactured from gum rosin, which comes from pine tree oil.

**Trihydroxystearin** is 1,2,3-propanetriyl tris(12-hydroxyocta-decanoate), melting point 86 °C. It acts as an emollient to sooth the skin and as a viscosity moderator.

**Tri-iodothyronine** is also known as liothyronine when used in pharmaceutical products. It is a hormone that regulates several metabolic functions, in particular the body's temperature.

**Trimethoxycaprylylsilane** is triethoxyoctylsilane and is used as a binder and emulsifier.

**3-(Trimethoxysilyl)propylamine** is an adhesion promoter; in other words it helps polymers to stick to shiny surfaces like glass.

**Trimethoxyvinylsilane** is $(CH_3O)_2C=C(OCH_3)SiH_3$. It helps to form cross-links between the polyurethane polymers and is known as an adhesion promoter.

**Trimethylolpropane** is 2-(hydroxymethyl)-2-ethylpropane-1,3-diol. It is used to modify polymers by cross-linking.

**Trimethylsiloxysilicate** is $(CH_3)_3SiOOSiO_2H$ and is used as an anti-foaming agent, emollient, and skin conditioner.

**Urea** is $(NH_2)_2CO$. It can act as a buffer, a skin conditioner, and as an antistatic agent. Urea is also known as diaminomethanal and carbamide.

**Vanilla extract** is made by grinding up vanilla pods and treating them with aqueous alcohol, which extracts the flavour molecule **vanillin**. However, most vanillin is manufactured.

**Vegetable oils** are also known as triglycerides and they consist of glycerol to which three long chain fatty acids are connected. The long chain fatty acids can be saturated (with no double bonds along the chain), monounsaturated (one double bond), or polyunsaturated (two or more).

**Viscosity improvers** (thickeners) – see Technical Words.

**Vitamin A** is an oil-soluble vitamin needed for growth, development, the eyes, and the immune system.

**Vitamin B$_2$** – see riboflavin.

**Vitamin B$_3$** – see niacin.

**Vitamin B$_6$** is also known as pyridoxine and is essential for making special proteins that the body needs. It is also involved in releasing energy from the body's store of glucose.

**Vitamin B$_{12}$** is also known as cobalamin, and it has a cobalt atom at its heart with a cyanide group (CN) attached. It is important for making red blood cells and the functioning of the nervous system.

**Vitamin C** is also known as ascorbic acid and has the food code E300. It is an antioxidant, and is important in wound healing and the maintenance of cartilage and bone. A deficiency of this vitamin leads to scurvy.

**Vitamin D** is needed to help the body absorb mineral nutrients like calcium.

**Vitamin E** is also known as alpha-tocopheryl acetate. It is an essential component of the human diet and is present in many plant oils, from which it is extracted. Vitamin E acts as an antioxidant.

**Water softener** – see Technical Words.

**Whey** is the liquid which remains when milk has been curdled, in the making of cheese. It contains mainly lactose, which consists of two simple carbohydrates, glucose and galactose, bonded to each other.

**White spirit** is a mixture of alkanes (hydrocarbons), distilled from crude oil. It is composed of alkanes containing 7–12 carbon atoms (chemical formula $C_nH_{2n+2}$) along with some aromatic hydrocarbons, although the benzene content must not exceed 0.1%. White spirit makes an excellent solvent for greases.

**Xanthan gum** is a complex carbohydrate consisting of linked rings of glucose, mannose, and glucuronic acid. It is made by the bacterium *Xanthanomonas campestris* fermenting corn starch. In food and cosmetic products it is used to increase the viscosity, and to stabilise a blend of ingredients and stop them from separating. It is also a skin conditioner. It has the food code E415.

**Xylene** is benzene with two methyl groups attached, of which there are three possible structures. It is used as a solvent.

**Zeolite** is a type of aluminium silicate with large cavities that can attract and hold calcium ions, thereby softening water. It is used in detergents.

**Zinc pyrithione** is a complex of zinc attached to two pyrithione groups. This is a biocide, effective against bacteria and fungi, and it will kill the fungus that causes conditions such as dandruff and athlete's foot.

**Zirconium hydroxide** is $Zr(OH)_4$. It is used in antiperspirants and deodorants.

# Sources and Web Sites

The ingredients in a product can often be obtained from its packaging. Sometimes the data is the minimum required by law and then the full information needs to be sourced from a company website. The major manufacturers are generally willing to reveal what a product contains and some even say why particular ingredients are used. The major manufacturers of household products are Unilever, Proctor & Gamble, Reckitt Benckiser, SC Johnson, Jeyes and Dr Beckmann. All have such a website although this is not always easy to locate or to find the ingredients in their products.

Unilever: http://www.unilever.com/innovation/Product-ingredient-safety/

Proctor & Gamble: http://www.info-pg.com or http://www.pg.com/productsafety

Reckitt Benckiser: http://www.rbeuroinfo.com/

SC Johnson: http://www.whatsinsidescjohnson.com/

Jeyes: http://www.jeyesprofessional.co.uk

Dr Beckmann: http://www.dr-beckmann.co.uk/products/ (very few product ingredients are listed)

Some websites, especially those of companies that sell online, such as supermarkets offering home delivery, also provide a complete list of ingredients for all the products they supply.

Chemistry at Home: Exploring the Ingredients in Everyday Products
By John Emsley
© John Emsley, 2015
Published by the Royal Society of Chemistry, www.rsc.org

Some companies do not want to reveal what their product contains and when I enquired I was politely told that the information was 'commercially sensitive.' Then it is necessary to resort to the MSDS (material safety data sheets) site which has to, by law, list certain ingredients.

Having obtained a list of ingredients, or at least of the active ingredients, it may then be necessary to locate their chemical composition and this can be done either through Wikipedia or The Merck Index®,[†] 15th Edition, ed. M. J. O'Neil, Royal Society of Chemistry, Cambridge, 2013, and The Merck Index *Online*.

There are also some websites where information can be found about ingredients that might have safety implications. The UK Health and Safety Executive (http://www.hse.gov.uk/index.htm) can provide information about all kinds of risks in all kinds of occupations. The European Commission has various sites that can be accessed via http://ec.europa.eu and these include Cosmetics, Enterprise and Industry, *etc*. The US Environmental Protection Agency (EPA) is also a source of reliable information and can be reached at http://www.epa.gov where there are sites devoted to various topics. The International Nomenclature of Cosmetic Ingredients (INCI) Directory has a website that provides information about the chemicals in personal care products and what each one is used for. See http://www.specialchem4cosmetics. com/services/inci/index.aspx#. The US Department of Health & Human Services (http://householdproducts.nlm.nih.gov) has a Household Products Database where information about household products, auto products, garden chemicals, personal care, and such can be accessed, and while they relate to products available in the USA, some are the same as those in the UK.

A lot of information about ingredients can be obtained from articles in magazines such as *Chemical & Engineering News,* the weekly magazine of the American Chemical Society, *Chemistry & Industry,* the monthly magazine of the Society of Chemical Industry, and *Chemistry World,* the monthly magazine of the Royal Society of Chemistry. The magazines *Education in Chemistry* (Royal Society of Chemistry) and *Chem Matters*

---

[†]The name THE MERCK INDEX is owned by Merck Sharp & Dohme Corp., a subsidiary of Merck & Co., Inc., Whitehouse Station, N.J., U.S.A., and is licensed to The Royal Society of Chemistry for use in the U.S.A. and Canada.

(American Chemical Society) also carry informative articles aimed at those teaching chemistry and their pupils. The Royal Society of Chemistry also publishes reports which address the various issues related to chemistry as well as a series of authoritative books on many topics. These are listed on the Royal Society of Chemistry website (http://www.rsc.org).

While a great deal of information can be obtained from the internet, books are more reliable because they have been checked by several people and been subject to review. (The same is true of the websites of official organisations.) Here are some books that I found useful.

*Essential Guide to Food Additives*, ed. M. Saltmarsh, RSC Publishing, Cambridge, 4th edn, 2013.

H.-D. Belitz, W. Grosch and P. Schieberle, *Food Chemistry*, Springer, Berlin, 5th edn, 2004.

T. P. Coultate, *Food: The Chemistry of its Components*, Royal Society of Chemistry, Cambridge, 5th edn, 2008.

*BMA Guide to Medicines & Drugs,* ed. J. Henry, Dorling Kindersley, London, 1993.

R. J. Kutsky, *Handbook of Vitamins, Minerals and Hormones*, Van Nostrand Reinhold, New York, 2nd edn, 1981.

*Pills, Potions, and Poisons: how drugs work,* Oxford University Press, Oxford, 2000.

T. Stone and G. Darlington, *Natural Toxicants in Food*, ed. D. H. Watson, Ellis Horwood/VCH, Weinheim, 1987.

G. Gardner, *Food Additives*, The Patent Office, London and Newport, 1992.

*Sixty-ninth Report of the Joint FAO/WHO Expert Committee on Food Additives*, World Health Organization Technical Report Series, no. 952, 2008.

The Royal Society of Chemistry publishes books that cover the chemical sciences and it also has a website http://www.rsc.org/ScienceAndTechnology/Policy/ where you can access papers on relevant topics such as the following:

*Proposal for a European Regulation on Novel Foods*, June 2008.
*Why Do We Worry About Phthalates?* December 2008.
*The Vital Ingredient: Chemical Science and Engineering for Sustainable Food*, January 2009.
*Why Do We Worry About Chemicals?* April 2009.

The Food Additives Industry Association (FAIA) publishes booklets such as *The Chemistry on Your Table* and *In the Mix*, which are designed to explain the need for food additives such as preservatives. Their website can be accessed at http://www.faia.org.uk/ and has a wealth of information about all kinds of food additives.

Finally, there are the books that deal with chemicals as they impinge on everyday life, namely the following:

P. Ball, *The Ingredients*, Oxford University Press, Oxford, 2002.

J. Schwarcz, *Let Them Eat Flax*, ECW Press, Toronto, 2005.

J. Schwarcz, *An Apple A Day: The Myths, Misconceptions, and Truths About the Foods We Eat*, Other Press, New York, 2009.

B. Selinger, *Chemistry in the Market Place*, Harcourt Brace Jovanovich, Sydney, Australia, 5th edn, 1998.

C. H. Snyder, *The Extraordinary Chemistry of Ordinary Things*, John Wiley & Sons, Inc., New York, 1992.

S. Quellen Field, *Why There's Antifreeze in Your Toothpaste: The Chemistry of Household Ingredients*, Chicago Review Press, Chicago, 2008.

# Subject Index